Urban Quay Walls

Maritime structures

CRC Press
Taylor & Francis Group
Boca Raton London New York

CRC Press is an imprint of the
Taylor & Francis Group, an **informa** business

A BALKEMA BOOK

Urban Quay Walls

Published by SBRCURnet

Published by:
CRC Press/Balkema
Schipholweg 107C, 2316 XC Leiden, The Netherlands

First issued in paperback 2022

© 2016 by Taylor & Francis Group, LLC
CRC Press/Balkema is an imprint of the Taylor & Francis Group, an informa business

No claim to original U.S. Government works

ISBN 13: 978-0-367-73733-7 (pbk)
ISBN 13: 978-90-5367-615-8 (hbk)

DOI: 10.1201/9780138734572

Liability

Editors:	MSc. A.A. Roubos
	MSc D. Grotegoed
Publisher:	Joshua Tourrich
Translation:	Textwerk, Amsterdam & Aad van den Thoorn
Printing & handling:	Printsupport4U, Meppel
Book & cover design:	Linda de Haan

CONTENT

Foreword

A large number of municipalities within the Netherlands have quay walls alongside canals and rivers. These quay walls are often old structures which, in many cases, are considered cultural heritage. There is no single overarching safety philosophy for these small-scale urban quay walls and the surrounding infrastructure. It is notable that, according to the assessment of certain quay walls, some collapsing should have already occurred whereas, in practice, the quay walls are still functioning despite dealing with far greater loads compared to the loads these walls were originally designed for. Clearly, the current design models do not provide a proper approximation of current safety.

Additionally, there is a lot of uncertainty about the management and maintenance of urban quay walls, where – in general – every municipality develops its own policy regarding management and maintenance. A request has arisen from within the sector to compile all this knowledge and experience regarding safety aspects, design, construction, and management and maintenance. This handbook has been compiled in accordance with the safety approach set out in the revised Handbook Quay Walls, CUR publication 211 Quay Walls.

It has been decided to make knowledge and experience with regard to the urban quay walls accessible and in doing so, to offer guidelines to all parties involved with regard to design, construction, management and maintenance.

The handbook is primarily intended for managers of municipal quay walls, for designers and implementers at commissioners, consultants and contractors.

CUR committee C 186 has supervised the development of this handbook. At the time of publication of this handbook, the composition of the committee was as follows:
- Ph.D. J.G. de Gijt, chairman, Municipality of Rotterdam/Technical University Delft
- A. Oskam, secretary, Municipality of Rotterdam
- MSc A.A. Roubos, rapporteur, Municipality of Rotterdam
- MSc D. Grotegoed, rapporteur, Ballast Nedam Engineering
- M. Bruchner, Engineering Agency Amsterdam
- J.C.M. van der Burg, Municipality of Delft, sector engineering agency
- MSc M.G.A. van den Elzen, Grontmij
- MSc P. Haltenhof, Engineering Agency The Hague
- Ph.D. R.K.W.M. Klaassen, SHR (on behalf of branch association F30)
- MSc H.A. van Laar, Hakkers (on behalf of VAGWW)
- M. Lodema, Nebest
- MSc P.J.M. den Nijs, Wareco (on behalf of branch association F30)
- MSc M.G.J.M. Peters, Grontmij

- MSc B. Rijneveld, Fugro GeoServices
- BSc M.M. van Veen, Municipality of Utrecht
- BSc L.C. van der Velde, Municipality of Rotterdam
- J.C. van Vliet, Municipality of 's-Hertogenbosch
- J.J. de Vries, Municipality of Haarlem
- MSc L.W.A. Zwang, Fugro GeoServices
- BSc P.A.A. van den Eijnden, corr. member, Engineering agency Drechtsteden
- BSc A. Jonker, coordinator, SBRCURnet

The editing of the handbook was done by MSc A.A. Roubos (Public Works Department Rotterdam) and MSc D. Grotegoed (Ballast Nedam Engineering). Special thanks to prof.ir. A.F. van Tol (TUDelft), prof.drs.ir. J.K. Vrijling (TUDelft), Prof.A.C.W.M. Vrouwenvelder (TUDelft), and MSc J. Kopinga (WUR) for their contributions.

Financial contributions for the realization of this handbook were received from:
- Municipality of Rotterdam
- City Management Services The Hague
- Municipality of 's-Hertogenbosch
- Municipality of Amsterdam IJburg and Zeeburgereiland
- Municipality of Delft, sector engineers
- Municipality of Haarlem
- Municipality of Nijmegen
- Municipality of Utrecht
- Grontmij
- Engineering agency Amsterdam
- Fugro GeoServices
- Nebest
- FCO GWW
- VAGWW

SBRCURnet would like to thank these institutions, as well as the committee members, who worked with great effort and enthusiasm on the production of this handbook.

Fred Jonker

Programme Manager Geotechnology and Soil

Summary

Since the thirteenth century, quay walls of significant retaining height have been built in the Netherlands in urban environments. Wood and stone were mainly utilised for these first quay walls.

Over time, structural revisions were often carried out due to the function no longer being fulfilled due to excessive deformations or the use of increasingly larger and deeper vessels. These historic quay walls were seldom designed for the current functional requirements they are subjected to. These mostly older quay walls require more maintenance and revisions. In some cases they need to be demolished and completely rebuilt.

The aim of this SBRCURnet publication is to present a proposal for a uniform and standardised method to verify the actual state of the current quay walls in urban environments. This report presents an overview of the development of urban quay walls over the course of time and describes the main types in use in the Netherlands. Subsequently, a method is prescribed for deciding whether or not a historic quay wall needs to be renovated, renewed or maintained. Within this decision making process aspects such as history of use, inspection, environment and technical boundary conditions are taken into consideration.

The actions necessary are determined by the quay's functional requirements. If these functional requirements can be fulfilled by sufficient control measures, structural adjustments may not be necessary. This report, therefore, takes an in-depth look at the management and maintenance of urban quay walls, and provides explanation on the collection and maintenance of up-to-date information regarding the current state of the structures, as an integral part of asset management. When an inventorisation provides insufficient data regarding the current state of an urban quay wall, an inspection of structural parts can provide additional information. This publication contains a separate chapter covering the inspection process. By providing a solid overview of the current state of an existing quay wall, this report can be used to decide whether and which measures must be taken. A method is also proposed for renovations and the construction of new quay walls, in which a specific safety philosophy is prescribed.

1 Introduction

1.1 In general

For centuries, quay walls of significant retaining height have been built in urban areas in the Netherlands. Wood and stone were mainly utilised for the first quay walls. Over time, structural revisions were often carried out due to the function no longer being fulfilled due to excessive deformations or the use of increasingly larger vessels. This publication provides an overview of the development of urban quay walls over time and describes the main forms of urban quay walls as occurring in the Netherlands. Subsequently, the steps for assessment regarding the need to adjust, renovate, or retain an existing quay wall are described. Aspects such as the history of the use, inspection, surroundings, environment, and the technical boundary conditions are discussed. The analysis of the technical boundary conditions and the assessment of the structure are linked to the results of the current SBRCURnet publications as much as possible.

Between 1990 and 2005, "Handboek Damwandconstructies" and the Handbook Quay Walls were published. "Handboek Damwandconstructies" primarily concerns the quay walls in excavations while the Handbook Quay Walls describes the design and construction of quay walls in sea ports. Specific aspects of urban quay walls, such as the influence of trees directly alongside the quay and handling during the remaining lifetime, are not discussed in the existing publications. Both books are used in the Netherlands as boundary conditions for registration as well as in education. These books, however, contain different design philosophies, which is undesirable within the framework of comparable designs for tenders. Therefore, in 2009, the CUR committee took the initiative to revise the Handbook Quay Walls and to harmonise the design philosophy. During the committee meetings, the idea arose to produce a Handbook Urban Quay Walls, as the Netherlands has many old urban quay walls. These quay walls were not designed for their present day use. Due to their old age, the structures require a lot of maintenance, revisions have to be made, or the entire quay wall has to be replaced by a new structure. This SBRCURnet publication aims to offer guidelines for a uniform assessment of existing urban quay walls in particular.

However, before moving to one of the measures, research must be conducted. The necessary measures are determined by the functional requirements of the quay. In Chapter 1, the various functions will be explained. When the functional requirements can be fulfilled by sufficient control measures, revision may not be necessary. Therefore, Chapter 2 discusses the management and maintenance of urban quay walls. Furthermore, this chapter elaborates on collecting and maintaining up-to-date information regarding the condition of the quay as an integral part of the management process. When inventorization provides insufficient information on the current condition of an urban quay wall, an inspection of structural parts may provide further insight.

Chapter 3 discusses the inspection in detail. A good understanding of the current condition of the structure can subsequently be utilised to test whether and which measures must be taken. This assessment method is described in Chapter 4. Repairs or new construction are referred to in Chapter 5, which provides a specific safety description. Finally, Chapter 6 provides a complete overview of the aspects of the implementation of the measures to be taken.

1.2   Functional requirements

Many of the urban quay walls in the Netherlands are old. The first quay walls in cities date back to the 13th and 14th centuries. Over time, these quay walls often got different functions, whereby the loads could be larger than the loads that were originally intended for the quay walls. As a result of changing functional requirements and the availability of the new materials and techniques, the shape of the quay walls in cities has also evolved. An overview of the development of quays in urban areas is shown in figure 1-1.

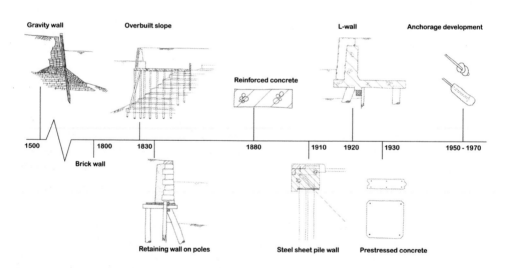

Figure 1-1 Development of urban quay walls.

A change in the quay's functional requirements often results in a less favourable load situation, for example due to an increased retaining height or increased top load. On the other hand, the function of an existing quay wall can also undergo positive changes whereby the load has decreased over the years. To be able to assess the possible measures to be taken, it is

12

therefore important to investigate the functional requirements of the current quay wall. This chapter provides an overview of the possible functions fulfilled by urban quay walls.

The various functions of an urban quay wall are:
- Retaining function
- Load-bearing function
- Mooring function
- Traffic function
- Storage function
- Environment function

Retaining function

The structure must be able to withstand the load from soil pressure and water pressure at a certain safety level. The retaining height is a determining factor, as is the difference between the top of the quay and the design depth. The design depth is determined by the depth of ships, if possible in urban waters, or by the amount of water to be drained.

Load-bearing function

The quay must be able to safely bear vertical loads from traffic, storage, and temporary supporting structures to the subsoil. Temporary loads, such as supply trucks that load and unload goods, may also be important.

Mooring function

When mooring ships to the quay wall, the quay must be able to absorb impact and hawser loads. Mooring must be able to occur safely. The designer must take into account additional loads arising from wind and currents. Waves generally do not play a significant role. Attention must be paid to the mooring space and possible passing ships. It is noted that quay walls within an urban area often have a limited nautical function. The emphasis mainly lies on recreational use.

Traffic function

Many roads in city centres have been constructed directly alongside quays. This means that the quay wall must also be able to withstand the loads from traffic and possible mobile cranes and revert this to the subsoil.

Storage function

Quay walls can be used for the storage and transshipment of goods. For this function, the substructure of the quay must be able to withstand the top load arising from the weight of goods that are stored.

Figure 1-2 Traffic function of an urban quay wall.

Environment function

Urban quay walls are located in an environment where aspects such as recreation, monumental value, zoning plans and zoning plan alterations must be taken into account.

1.3   Main forms

Figure 1-1 shows that not only has the function of an urban quay changed over the years, but the structural building as well. Moreover, this report will distinguish between four different types of quay walls:
- type 1: gravity wall on spread foundations
- type 2: gravity wall on pile foundations
- type 3: L-wall on pile foundations
- type 4: Sheet pile wall in steel or concrete

Section 3.2 will discuss the inspection methods based on this categorisation. In section 4.3.1, the test method for each main type is discussed in detail while, Chapter 6 uses the categorisation into main types for the safety philosophy and accompanying safety factors.

2 Management, maintenance and inventory

2.1 Introduction

This chapter discusses the technical management and maintenance of quay walls in urban areas. A manager will frequently inspect the condition of the quay structures in their area. In fact, the manager is constantly collecting (taking inventory) and gathering information (inspecting). In order to be able to manage, it must first be clarified which structures are included in the manager's area. Subsequently, the manager must have a clear overview of the functions and allowed loads. This information helps the manager to guarantee the safety and the functioning of the structures in the management area and to anticipate possible alterations in function or load. During the management phase, the manager must be able to answer the questions following:

- Which functions does a structure have and which requirements must the structure meet?
- Is all technical information of the basic geometrics known and has it been collected?
- What are the critical components of the structure, which risks are involved, is the structure still safe?
- What is the maintenance condition of the structure?
- When must the structure be repaired/replaced?

As is also emphasised in the handbook Quay Walls [8], it is important that the manager takes a proactive approach. The manager must guarantee that the functional terms of reference are met. During the design of a new structure or the repair or renovation of an existing structure, the manager's quality objectives must be specified and drawn up. If the manager clearly formulates the acceptance criteria, this will facilitate and simplify the transfer of a new structure.

When drawing up the quality objectives, the quay walls are schematised and possible threats to these objectives are analysed, as well as the possible influence of aging and overloading.

Figure 2-1 Rotterdam, Aelbrechtskolk historical Delfshaven.

2.1.1 Asset management

The management of structures is also referred to as asset management. Asset management is a method in which the activities take place in a systematic and coordinated manner. Consequently, an organisation can optimally manage its assets (objects), performances, and expenses for the entire lifetime in order to achieve the organisation's strategic objectives.

By maintaining the objects, a contribution is made to the accessibility and the quality of the public space in the inner city.

Implementing asset management endeavours to place greater emphasis on the balance between risk, performance, and cost. Decisions regarding planning and measures will be taken from a Life-Cycle-Cost approach.

It is important that the different roles are distinguished. In trade jargon, these are the asset owner (board), the asset manager (manager), and the service provider (execution). Where the asset owner determines the vision and objectives, the asset manager determines what must be done based on the vision and objectives, and the service provider handles the realisation of the construction.

2.1.2   Technical management

Technical management of quay walls is the determination, in consultation with all disciplines involved, of all measures needed to ensure that the objects can retain or achieve their agreed upon functions and qualities for a longer period. This is done by creating conditions and recording agreements reached, as well as performing all tasks assigned to the technical management by the board. Management is the actual execution of the activities arising.

The management and maintenance policy is aimed at maintaining the structure and retaining the functionality of the object during its lifetime. The lifespan can be extended as long as this is structurally and economically responsible. Safety is paramount and must always be guaranteed. In principle, technical management relates to the unaltered conservation of the object.

As mentioned previously, the manager must have insight into the complete area. The circle diagram below, figure 2-2, presents the urban quay walls in Rotterdam. It is clear that this is an area with a relatively high number of urban quay walls.

Therefore, the manager must be attuned to this. Old quay walls display relatively more flaws. Consequently, it may be necessary to monitor and inspect these more frequently. Furthermore, more frequent investigation to repair damages may be required. The frequency is determined by a risk analysis that provides insight into the risks and chances of failure.

When the risk analysis shows that a component of the quay wall no longer suffices, it must be investigated whether the lifespan of the quay wall can be extended in an innovative manner. In short, an area with old quay walls requires more intensive management than an area with newer quay walls.

In other older Dutch cities, the ratio is most likely to be dominated by the categories 51 to 80 and 81 to 100 years old.

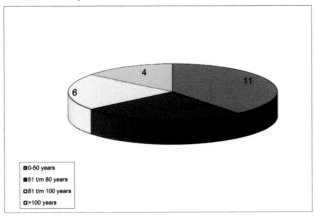

Figure 2-2 Area quay walls, categorised by age, in Rotterdam.

2.1.3   *Functional management*

The functional management is responsible for the allocation of use (contract forming) and the inspection of the compliance of prescribed conditions and rules. The manager must take the environmental factors and requirements of the quay structure into account.

In the past, waterways mainly had a nautical function, whereas nowadays, the urban waterways may also fulfil a recreational function in addition to the nautical function. The many kilometres of canals have gradually changed function over the course of the decades. The industry of the past has disappeared in many ports and canals, or has been replaced by new market segments, such as party ships, tours and cruises, water taxis, public water transport, recreational/leisure boating, events, and permanent mooring (museum boats, houseboats, restaurants/hotel boats). It could be said that nowadays, the quay walls are used/viewed from the 'land' more than they are from the 'water'.

Functional alterations are included in the functional management, but can certainly influence the technical management. Therefore, it is important that proper coordination takes place within an organization regarding the functional management and the accompanying technical conditions. Where the aforementioned functions have been separated, the functional management is responsible for maintaining the collaboration between the main user on the one hand and the technical manager on the other hand.

Upon drawing up an integral vision for urban management, where the terms of reference plays an important role, the following aspects may be of importance:
- Reserving space for future initiatives.
- Public access quays.
- Diversity of vessels.
- Nautical functions.
- Recreational/leisure functions.
- Use of the ports and quays matching the profile and the environment.
- Distinctive ports.
- As much liveliness as possible in the centre.
- Water-related events in the centre as far as possible.
- Parking/traffic safety.
- Waterway management.
- Facilitating (market) initiatives.

It must be noted that the above aspects will differ per municipality and must therefore be determined per municipality in order to perform functional management.

2.2   Management philosophy

In order to create a well-substantiated management plan, it is necessary to get an accurate image of the functional requirements for the objects to be managed. In other words: which functionalities can be attributed to the urban waterways, which plans exist for future zoning of the waterways, and on which maintenance levels must these waterways be maintained? These ambitions can differ per port or even per quay.

2.2.1   Quality ambitions

The quality level of the quay walls may be linked to the maintenance level of the adjacent public space. Various policy pieces can provide input for the forming of a management vision:
- Spatial development strategy.
- Policy documents.
- Water plans.
- Management and maintenance public space.
- Urban ambitions (city vision).

In addition to national norms and guidelines, municipal policy must also be taken into account. The quality ambitions can be depicted per municipality. Figure 2-3 presents the quality ambitions in Rotterdam's exterior space. Red and green indicate areas with an 'extra' quality level, while the remaining parts are 'standard'. As the figure shows, the majority of the Rotterdam quay walls in the city center border exterior space with a high quality level. The quality framework for quay walls is formed by the aspects 'safety', 'functionality', and 'presentability'.

The definition of these aspects is explained below:

Safety: a quay wall and components of this wall must be of such quality that the chance of personal injury and economic damage as a result of collapsing or insufficient functioning is kept to a minimum.

Functionality: a quay wall (and its components) must be of such quality that it meets the requirements of the current use. This means that the quay must be intact; possible small areas of damage are allowed, if these do not have a negative effect on the quality of the structure or use of the quay.

Presentability or appearance: in principle, the aesthetic requirements for a quay wall are subjective and are related to the location of the quay wall in the municipality and the recreational functions and functions of the space above the quay wall.

Maintenance measures are executed when one or more of the abovementioned requirements are not met. It must be determined whether the investment for the repair works can be justified with regard to the intended lifespan of the structure. In certain cases, full replacement or thorough renovation may be required.

Figure 2-3 Urban quality map Rotterdam.

2.2.2 *Rational management*

Effective and planned construction of preventive maintenance is the basis for the method of rational management. Rational management means activities that form a coherent whole, which are executed to maintain the quality of the area at the lowest possible cost.

This method has significant advantages compared to a method where action is only taken in case of disruptions and/or issues with maintenance. These advantages are:

- Decommissioning of objects occurs in a methodical and predictable manner.
- Activities can be optimally attuned, also with regard to traffic disruptions and emergency services.
- The risk of accidents is reduced.
- No consequential damage.
- Longer lifetime.

2.2.3   Integral management

In many cases, the function of a quay wall is not limited to merely water and earth retaining functions. Quay walls also have an urban developmental, cultural-historical and/or ecological value.

Important secondary functions of quays are functions such as walking and recreational activities. Additionally, a quay can contain cycling paths, driving lanes, or parking spaces.

In the execution of the management tasks, other functions than the primary function of the quay wall must be taken into account.

The management of a quay wall is integral: care for the water and soil retaining function with an eye on the interest arising from the user functions. Integral management is also expressed in the coordination of various projects in the planning phase. Double use of quays (or the area within the sphere of influence of the quay walls) can arise due to a change of function or redevelopment of an area.

2.2.4   Planned management

Planned management is important to obtain insight into the quality condition and quality development of the elements to be managed. With planned management, maintenance measures can be planned at an early stage. Consequently, the quay wall can be maintained at the lowest possible cost during the technical lifetime.

Planned management means that:
- Quality inspections are executed in accordance with established criteria.
- Quality inspections form the input for the determination of the residual lifetime.
- Residual strength and lifespan of quay walls are determined at least once every 10 years.
- Inspections take place in accordance with a multi-year program.
- A system with quantity and quality information is available.
- Maintenance requirements are determined on the basis of established quality criteria and residual lifespan.
- Based on inspections, the maintenance requirements can be adjusted.
- A valid multi-year maintenance program, including financial substantiation, is always available.

2.2.5 Safe management

The management strategy in urban areas is aimed at preventing failure of the structures. From a safety perspective, it is irresponsible to delay maintenance until a quay wall no longer meets the safety requirements. This principle forms the basis for ensuring durable safety of the quay walls.

Safe quay walls meet the following requirements:
- The quay walls meet the applicable safety norms regarding legislation and regulations.
- No unacceptable subsidence may occur and no holes may arise behind the quay walls.
- The quay walls must not display any structural flaws or defects.
- The quay walls are in such a structural state, and the maintenance and management are performed in such a way, that protection against collapsing is safeguarded.
- Additional functions must not have a disadvantageous effect on the water and ground retaining capacity of the quay wall and must meet the functional terms of reference.

Existing quay walls increasingly fail to meet the calculation rules in the valid structural codes. This can be the result of changing functionality and overloading, degradation of the structure, new structural guidelines, or changing calculation methods.

If inspections demonstrate that the current load on quay walls is higher (for instance due to increased traffic load) or structures have significantly deteriorated, recalculations are then performed in accordance with the valid structural norms. The backgrounds for such an assessment are described in further detail in Chapter 4. The board is advised regarding the measures to be taken on the basis of the assessment results.

2.3   Management tasks

In order to realise an efficient management of the quay walls, various subtasks must be performed. The tasks belonging to technical management are described below. The management is primarily responsible for periodic and major maintenance, replacement investments and failure maintenance.

2.3.1 Advising

One of the main management tasks is advising the board regarding the layout, use, management and maintenance of the quay walls. In the form of maintenance plans, multi-year plans and annual plans, the manager makes proposals regarding the maintenance to be performed to ensure that the quay walls meet the agreed quality level. This includes the substantiation of the financial need for management and maintenance in the short and long term. It is important to take into account the requirements and conditions of the board, insofar as this is possible, as well as the wishes of citizens and companies as included in the functional terms of reference.

There is also coordination with projects initiated by other managers, investors and owners. Participation in consultations concerning restructuring plans, area developments, zoning plans, etc. also takes place.

Management plan

The main objectives of a management plan are:
- Maintaining a proper functional division between land and water in a way which is friendly towards nature, the environment and the surrounding area, with an efficient utilisation of means.
- Further professionalising of management and maintenance tasks.
- Preservation and recovery of cultural values.

The management plan provides insight into the planned maintenance and the framework of management and maintenance costs. It describes the technical measures needed for the long-term functional maintenance of quay walls. Additionally, this gives other parties insight into the policy, such as the province, the water boards, individuals, companies and other governments. This can contribute to the effective coordination of activities and projects.

The management of quay walls is connected to a number of other maintenance measures such as:
- Management of the water and bottom in connection with water quality, fishing up rubbish and dredging activities, for instance.
- Management of elements that are separate from the quay walls, such as mooring poles, mooring fixtures and shore power boxes.
- Management and maintenance of berths and piers, gangplanks and other objects.
- Nautical waterway management.
- Green management.
- Maintenance in the framework of Flora and Fauna, as prescribed in legislation and regulations.

Terms of reference

The manager draws up a Terms of Reference for large-scale maintenance projects and replacement which includes the functional and technical requirements for the new quay wall.

2.3.2 Damage settlement

Management tasks also include the settlement of damage reports regarding core tasks, consultation with reporters and ensuring completion of damage or customer settlements. Furthermore, management tasks include answering questions and wishes from citizens. It is important that the service norms are met. It must be stated within how many days damage reports must be settled or the settlement process must commence. Most damage reports are filed as a result of a visual inspection, via road authorities, or users/residents.

2.3.3 File management

For the management process, it is important to be able to have access to current object information, both regarding administrative data, data concerning the size of the area (also on a detailed level) and regarding periodic inspections. File management also includes adding information and keeping it up-to-date in a digital management system.

File management strives to gather at least the object information below:
- location: district, neighbourhood, sub-neighbourhood;
- ownership;
- object information/decomposition: length/surface, date of structure, type of structure, foundation, dimensions;
- basic cross-section;
- materials: stones, brickwork, natural stone, concrete, wood;
- performed maintenance: minor maintenance, partial maintenance, reconstruction, jointing, painting;
- inspection information: levelness, irregularities, cracks, damages;
- inspections: condition measurement, material-specific inspections;
- measurements: deformation measurements, levelling, load capacity;
- pictures;
- terms of reference: maximum allowed loads and nautical use;
- geotechnical information;
- calculations and test results;
- historical data;

If data is missing, it is difficult for the manager to give correct repair advice in case of damages. In those cases, the only solution is often to perform a detailed and relatively expensive investigation, for instance by partially digging out the structure. Managers who have insufficient information regarding the basic geometrics of the structure in the area, can conduct a class 0 inspection. Also see Chapter 3.

The advantage of digitalising information and data in a management and information system is that the information is accessible for multiple parties and does not have to be sought in various systems and archives.

A completion file must be provided to the manager immediately after completion of a new development or renovation project. This file must consist of:

- statement of inspection;
- final settlement;
- transfer form, including:
  - project evaluation, both technical and financial;
  - tender specifications and information notice;
  - work specifications;
  - specification drawings;
  - construction calculations, including: starting points (top) load and geometrics;
  - registration bill and statement;
  - order letter;
  - construction reports;
  - quality documents (certificates/inspection forms/guarantee statements);
  - measurement grounds (baseline measurement);
  - deviation forms;
  - list with key ratios;
  - as built drawings / revision drawings;
  - pictures;
- updated version of the overview drawing and the cross-sections for the management system.

This information must be processed in the management system, to ensure that the most recent information is available in case of damage and maintenance.

2.3.4 Inspecting and testing

Regular testing is an important and vital means to maintain the quay walls in a sustainable and safe way. Chapter 4 discusses the assessment of an existing structure in order to guarantee the minimal safety level in more detail. It is noted that a clear relationship exists between testing and inspecting. Insight into the current state of the area is gained by conducting inspections of the structures. Also see figure 2-4 for a complete maintenance process description.

Based on the data from the inspection and the condition measurements, it is assessed whether the development or aging of the objects matches the theoretical residual lifetime and it is determined whether the planned measures must be performed in accordance with the plans. It can also be determined whether additional or accelerated measures are needed to repair the deviations which have been discovered. The inspection methods and classes are described in further detail in Chapter 4. This information can be used for making a request for inspections.

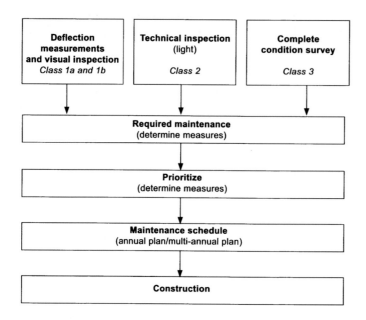

Figure 2-4 Process description management urban quay wall.

2.4 Maintenance

Maintenance can generally be subdivided into a number of forms as shown in figure 2-5. The maintenance stated here consists of repairs and inspections.

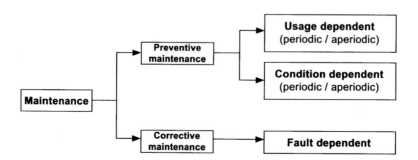

Table 2-5 Types of maintenance.

The difference between preventive and corrective maintenance is characterised by the failure limit of the structure's component. Failure means that the structure's component no longer meets the functionality requirements, due to which corrective action is necessary. The actions to be taken for preventive and corrective maintenance are considered in a risk analysis.

26

This action is performed after failure of the structure's component, with the goal of repairing the required functionality and safety. The term 'corrective' relates to the repair of the functionality to the desired level.

The aim of applying a preventive maintenance strategy is to conduct maintenance action before the structure component fails. This means that damage can be repaired before it becomes so large that it endangers the functionality of the structure and leads to failure. One reason to opt for a form of preventive maintenance could be that failure would have disastrous consequences. In that case, the allowable risk of failure is small.

Preventive maintenance can be subdivided into:

Usage dependent maintenance: In this form, a component of the structure will be repaired after a fixed number of usage units, irrespective of the state. The fixed number of usage units gives rise to a further subdivision in periodic usage maintenance and non-periodic maintenance.

Condition dependent maintenance: This form of maintenance has a stochastic character in time. Before progressing to repair or replacement, the component concerned is first inspected. Depending on the state of repair and the expected maintenance work, the decision is made whether or not repair is needed and when the next inspection must take place. This form can also be distinguished from periodic condition-dependent maintenance (such as yearly inspection of the structure) and non-periodic condition-dependent maintenance.

In practice, maintenance to quay walls is mainly of a preventive nature. The required criteria are drawn up beforehand. Repair is required upon exceeding these criteria. The criteria stated must be significantly below the failure limit.

Corrective maintenance is possible when the chance of failure can be established relatively quickly. As this is difficult in the case of quay walls, especially for the components that are located under water, corrective maintenance is not an option. This could have disastrous consequences for the safety and availability of the structure.

The exact moment of failure is relatively difficult to predict. Condition-dependent maintenance is therefore often used in practice. The length of the inspection interval can be adjusted during the lifetime of the structure. In the beginning, during the design lifetime, the inspection interval is practically constant and can be adjusted after the design lifetime based on the degradation of the materials of the structure or change of usage.

Maintenance to quay walls can be divided into disruption maintenance, periodic maintenance, and project-based maintenance. In this section, the activities belonging to each of these types of maintenance will be discussed in further detail.

2.4.1 Failure maintenance

The primary objective of management is to perform preventive maintenance on the quay walls. Nevertheless, unexpected damages may sometimes arise that acutely inhibit the functioning of a quay structure. These flaws or damages usually have an external cause, such as a berthing forces, construction activities, too much traffic load, or leaking pipes.

Deviations can also arise within the structure itself, such as:
- Accelerated degradation of wooden foundation and floors, for instance due to dry periods, mould, and bacteria.
- Accelerated degradation of steel due to corrosion.
- Leaking ground retaining screen (a steel quay wall or wooden quay wall).
- Non-functioning drainage.
- Dilatation joints and connections with adjacent structures.
- Cracks, structural or non-structural.
- Soil erosion in the water bottom.

Failure maintenance consists of the measures needed to repair these damages or deviations and by doing so, reinstating the functionality of the quay wall. Examples of this are included in the figures below and on the next page.

Figure 2-6 Repair dilatation joint.

Figure 2-7 Subsidence due to under-seepage.

Figure 2-8 Damaged upright course of bricks by storm.

Figure 2-9 Subsidence.

2.4.2 Periodic maintenance

With these measurements, the quality level of the quay wall is maintained at the required level as efficiently as possible. This mainly concerns:

- Weed control on capping beam and removing seedlings quay wall (unless it concerns protected plants).
- Spray-cleaning concrete and brickwork, this is an activity that must only be applied in moderation to avoid damage to the joints. Spray-cleaning may be necessary for representative parts of the ports. The flora and fauna must be taken into account. For inspections, it can also be decided to spray-clean if unseen damage is suspected.
- Cleaning drainage pots.
- Joint repair is necessary for many old stone quay walls. Only perform joint repair when inspection shows that the quay wall structure will remain operational for several more decades. Jointing is generally expensive as provisions must be made to work partly under water.
- Conserving the quay components (corrosion, bollards, steps, railings, etc.).
- Conserving the quay wall or applying cathodic protection.

Figure 2-10 Corrosion of a bollard.

Figure 2-11 Wear and tear joints basalt wall.

Table 1 provides an overview of the measures and maintenance intervals for periodic maintenance, that are related to the two quality levels. The frequency with which the maintenance measures are conducted is dependent upon the urban ambition level on location and the connected quality desired; standard or extra.

Table 1 gives an indication of the periodic maintenance measures for the urban quay walls of Rotterdam.

Table 1 Periodic maintenance measures with common frequency in Rotterdam.

Periodic maintenance work	Interval (in years)	
	Standard	Extra
Weed control capping beam	1	1
Removal seedlings quay wall	2	1
Spray-cleaning quay wall	6	3
Cleaning drainage pots	10	10
Joint restoration	15	10
Paintwork (bollards, steps, banisters, etc.)	8	4

2.4.3 Project-based maintenance

Project-based maintenance includes those maintenance activities that do not belong to 'periodic maintenance' and do not meet the criteria for 'replacement investments'. This concerns (large-scale) activities performed to ensure that the quay wall once again meets the norms applicable to quay walls.

Project-based maintenance to quay walls in an urban environment is complex. In order to determine the extent and nature of the optimum maintenance measures, a complete condition inspection (inspection class 3) is often conducted. If necessary, control calculations are made. This inspection is often supervised by specialised inspection agencies. The inspection results must be recorded in the management document. This ensures that new results can easily be compared with the previous inspection results. Strength calculations of structures can be made based on the information provided in the management document.

Project-based maintenance is often drastic for immediate residents and road users. In order to limit the nuisance, the project-based maintenance is:
- attuned to the maintenance of other objects such as roads, bridges and buildings in, on or near the quay wall as much as possible;
- performed in such a way that the period between two consecutive project-based maintenance measures is at least 10 years.

Project-based maintenance can be performed in the form of a renovation as well as restoration. In case of a renovation, a substantial part of the existing structure is applied in the new structure. In case of a restoration, the original design, applied materials and connection structures are used as much as possible.

The list below provides a number of examples of measures that are included in project-based maintenance:
- Repair of fendering damaged by dry rot.
- Closing dilatation joints.
- Repair of ground retaining screen.
- Repair of worn fascine mattress (usually constructed when ports are deepened).
- Repair of holes.
- Replacement of brickwork.
- Vertical and horizontal repair of cracks.
- Grouting capping beams.
- Settling damage.
- Replacement of old components.
- Applying screw injection anchors and steel walings.

2.5   Inventory

2.5.1   In general

During the management phase, the manager must collect and gather all information regarding an asset in the area. Recording the information may be done by means of file management. In practice, multiple parties are often involved in an inspection of a quay structure. It is possible that a municipality is the owner, the water board is responsible for the water retaining structure and the bottom level, the Department of Infrastructure & the Environment or a port service is responsible for the management of the waterway, etc.

It is vital to have insight into the parties involved when collecting the information. As archive documents are not automatically transferred or available, it is important to have insight into the parties that were involved in the past.

Examples of archive documents that are wise to include in the file management are:
*   Terms of reference (both old and current).
*   Tender specifications.
*   Revision drawings and pictures of construction.
*   Geotechnical information and groundwater levels (Basic Registration for Subsurface Data).
*   Inspection results (assessment, deformation measurements, condition measurements, damages).
*   Calculation reports (design and testing).
*   Listed building assessment.
*   Flora and Fauna.
*   Cables and pipes.
*   Bordering structures.

Gathering this information in a timely and transparent manner could save significantly on costs. Especially when the basic geometrics are unknown, there is a high level of uncertainty in the reliability and the safety of the structure. The execution of inspection class 0 is relatively intensive which makes the costs for this class relatively high. Another important aspect in the assessment and management of quay structures is to have insight into the allowable loads on the structures.

The specific loads for quay structures can be found in section 4.2. The loads often arise from the functionalities of a quay structure or the connection with the immediate surroundings.

During the lifetime of urban quay structures, possible alterations may occur in the functionality of the structures. Think of the planting of trees and the construction of roads in locations where storage and transhipment activities used to take place. Or a quay that is nowadays only used for water management or as a boulevard.

2.5.2 Terms of Reference

There is a significant chance that the existing terms of reference will deviate from the specifications that applied at the time of the construction of the historical quay structures. In the past, a distinction was often made between Functional Terms of Reference and Technical Terms of Reference (FToR and TToR). Nowadays, this division is no longer as strict and an analysis is even made of the availability and reliability of a structure, based on risk-oriented management and maintenance.

This is also referred to as a RAMS analysis. In setting up the terms of reference, it is common to subdivide the requirements into:
- Functional requirements.
- Aspect requirements (RAMS).
- Realisation requirements.
- Connection requirements.

The functional requirements concern primary and secondary functions of the structures. The aspect requirements represent the requirements set for availability, reliability, maintainability, design and safety. Limitations usually apply during the realisation or the execution of maintenance. These limitations are often project-specific and can be translated into realisation requirements.

In addition to the functional factors, one of the most important factors that sets boundary conditions for a structure is the surrounding environment. Residents and businesses, united in communities of interests, and residents associations and foundations, all set requirements regarding the appearance and safety of the quay wall. When a quay structure is chosen as a system, the influence of the environment on the system can be considered as an external connection point. Requirements applying to the urban quay structure may arise from these connections.

2.5.3 Critical structural components

Chapter 2 includes various types of quay walls. For the inventory, the critical structural components can be listed per type of quay wall and the possible failure mechanisms can be noted. Please refer to section 4.3.1 for more information.

2.5.4 Condition measurement, degradation, and damage

An inventory can be performed based on various inspection classes, by means of a condition measurement. Please refer to Chapter 3 for more information.

2.5.5 Listed building assessment

Some quay structures in urban areas are either part of the protected cityscape, are listed buildings or are connected to a listed building or bridge. Moreover, a municipal, provincial or national listed building may be present inside the area of influence of the quay structure. In such cases, coordination with the design review committee is necessary and this committee will impose requirements to ensure that the characteristic cityscape is maintained. This is a highly determining factor for the manner of construction during maintenance activities or replacement projects. In the case of renovation projects, for example, the reuse of materials may be required, in order to maintain the original character.

Depending on the listed building status, a deviating course of action must be taken in comparison to new quay structures. The valuation and preparation of repair activities of historical quay structures with a certain listed building value requires the necessary knowledge and experience. This deviating course of action is especially noticeable in the preparatory phase and the definition phase in the form of a restoration plan, often preceded by a feasibility study and historical building archive investigation.

One must be in possession of a listed building permit issued on the basis of a restoration plan for the construction. The runtime of such projects is much longer compared to the runtime of a standard course of action. The listed building permit may set additional requirements. The application period for obtaining such a permit at the Cultural Heritage Agency can be up to six months. A sound restoration plan minimises the chance of delay in obtaining the listed building permit. In addition to obtaining this permit, a subsidy application is often ongoing for restorations. The restoration can only start after the subsidy decision. All activities performed before a subsidy decision are not eligible for (partial) compensation.

The restoration of a quay structure must be combined with the current requirements for safety and function. Due to the obligation to preserve the object, this is often quite a challenge, also in financial terms. Restoration is generally more expensive than 'standard' repair or new development. An example of this is the reuse of old construction materials.

Feasibility study

A precise and accurate determination of the current condition is vital in order to substantiate the feasibility of a restoration, even more so than in the case of a quay structure without listed building status. Generally speaking, the same inspection method can be applied for the feasibility study as is the case for quay structures without listed building status. However, it must be deliberated per component whether the scope of the inspection is sufficient. The scope is dependent upon the historical value and the assigned status.

Historical building investigation

In order to arrive at the proper deliberation as to how the quay wall must be restored, it is vital to be well-informed of the history and meaning of the civil engineering structures. By means of archive investigation as much information as possible is collected: (specification) drawings, previously performed maintenance or adjustments and the historical and geographical context. The historical building archive investigation is especially broader with regard to the societal aspect than an archive investigation for a quay structure without listed building status. Think of old pictures from the municipal archives or the local history association. Furthermore, events such as collisions and incidents from World War II are of societal importance.

The historical building assessment follows the historical building archive investigation: the quay wall itself is also a valuable source of information about the history and the function. This can be combined with a class 3 inspection of the quay wall.

Restoration plan

A restoration plan must consist of at least the following components:
- A description of the listed building structure, preferably in combination with a historical building investigation.
- An assessment of the components of the listed building.
- A vision of the restoration and a description of the objective of the restoration.
- A description of the current condition on component level. In the description, the materials used, the use of colours and the processing method of materials are discussed.
- A description and account of the components of the chosen restoration methods.
- The function of the object after restoration: will the original usage be preserved or not. Moreover, starting points must be formulated regarding the listed building value, safe usage and functional requirements.
- Drawings, if possible with original documents, of the existing situation.
- A budget framework on component level.
- A description of the course of action with a timeline.

Assessment

In order to arrive at an objective assessment and subsequently make a responsible choice, it is advisable to use an assessment table, which incorporates all aspects with regard to the assessment. An example of such a table is shown in appendix 2.

Restoration vision

A restoration vision includes a description of the course of action based on the functional requirements. The fact that preservation is more important than replacing the structure is central. In general, it will be opted to conduct the renovation of a listed building quay structure in an old city centre as unobtrusively as possible. The most important reason for this is the preservation of the authentic character of the city centre. Restoring the current situation as a whole is not always desirable. Restoring the structure (partly) in accordance with a specific period in time, is also possible. However, there may also be reasons to give shape to the restoration in a creative and striking way. In all cases, it is essential that the restoration plan is well thought-out and matches the urban planning within the area. Consultation between all parties involved is therefore of vital importance.

2.5.6 *Flora and fauna*

In the maintenance performed on a quay structure, the flora which is present must be respected. It must be noted, however, that the integral safety of the quay walls may never be compromised as a result of the present flora. In order to meet the flora and fauna legislation, the protected wall plants on the brick walls must be inventoried by an acknowledged ecologist. If protected plants are found, so-called mitigating measures must be included in the report. These are measures that must be taken to save and preserve the plants during the repair or restoration. Furthermore, a monitoring plan is included to assess whether the plants have not been damaged due to the restoration. This must also be reported.

The execution of activities that involve protected plants sometimes requires a permit from the ministry of Agriculture, Nature and Food Quality as part of the execution of the General Conditions Environmental Rights Act. In addition to the 'red list', various municipalities have a supplementing list containing protected species. It is important to ask the environmental officer whether such a list applies. In addition to the application of the General Conditions Environmental Rights Act, an exemption obtained from the Ministry of Economic Affairs, Agriculture and Innovation is sometimes required in accordance with article 75 of the Flora and Fauna Act.

Figure 2-12 Multiple species of ferns on a quay wall.

Prior to possible work activities on quay structures in urban areas, it is necessary to take stock of the protected flora and fauna. On the quay walls themselves, this mainly concerns various species of ferns that flourish in limy joints, see figure 2-12. Characteristic of ferns is that they are not wood forming and for that reason are not damaging to the structure. Always check whether a list of protected flora and fauna exists and can be viewed. If the bottom has been disturbed or dredged in front of the quay structure, a fauna investigation will be necessary.

Prior to maintenance activities it must also be established whether or not the ecological values have been compromised. This is important in the case of activities such as spray cleaning quay walls. This is not allowed to be done in places where protected wall vegetation is growing. In case of renovation and restoration activities, additional protective measures must be taken if necessary. In case of repair and replacement, an adaptive structure and an adjusted construction method may be chosen.

Grasses and wood-like crops

Wood-forming crops are characterised by their eventual development into trees or bushes. Both grasses and wood-forming crops can damage the connecting materials (cement) between the elements of a quay wall. Consequently, the strength of the quay wall is diminished. Therefore, it is important that these types of crops are regularly removed. Disadvantageous consequences for protected flora must be prevented as much as possible during restoration work.

Moss

Moss may also be growing on concrete and brickwork. If this causes an aesthetic problem, the moss may be removed by means of a high-pressure spray.

Trees

When trees are located close to a quay, they can lead to instability of the quay structure. Historic quay structures are often not suited for pressure due to trees. From this point of view, trees immediately behind the quay wall are undesirable. In addition to having a water retaining function, a quay wall may have functions that require the presence of trees. In these cases, it must be considered whether trees can be planted according to certain boundary conditions. Therefore, this publication includes calculation rules and practical guidelines to safeguard the interaction between trees and the structure, please see section 4.2.4.

Practical example

At a reconstruction on the Mauritskade in The Hague, pieces of brickwork with plants were taken from the quay and kept together with clamping straps. These pieces were subsequently stored on pontoons with the same orientation as the wall and were regularly watered. The thickness of the brickwork was increased for moulding the brickwork with plants back into the structure, and the quay wall was fitted with recesses for the water supply.

2.5.7 Connecting structures

Quay walls are often connected to other engineering structures such as bridges, locks, parking garages and buildings. In many cases, it is necessary to establish a clear border in terms of ownership and management. The unbuilt space within the area of influence of a quay wall must preferably remain undeveloped. In case of a considerable (societal) interest, building within the area of influence of a quay wall can be allowed under certain conditions. This must be considered per case. The most important boundary condition for buildings within the area of influence of a quay wall is that they must be built in such a way that the quay wall can be repaired or replaced unhindered and without additional measures and without the buildings being damaged by repair or replacement activities.

2.5.8 Cables and pipes

Due to the limited space, it is unavoidable that cables and pipes are placed in the ground within the area of influence of quay walls. Extra attention must be paid to compaction when placing the cables and pipes. In the case of subsidence behind the quay wall, it must always first be investigated whether recent activities on cables and pipes were performed to prevent the need for expensive investigations on the quay wall. The presence of cables and pipes increases the costs of works on the existing quay wall. When designing a new quay, the location of the cables, pipes, sinkers and drillings must be taken into account.

Within the application procedure regarding the location of cables and pipes, the manager of a quay wall is asked to assess the risks involved in excavations within the area of influence of the quay walls, if applicable. Conditions may be set for the owner of the pipes in granting the permit. After placing cables and pipes or drillings, a revision of the location and the depth must be recorded.

Figure 2-13 Brickwork pushed away by the growth of roots.

2.6   Degradation and choice of building materials

2.6.1   Introduction

For urban quay walls, materials are used in conformity with the requirements and regulations of the Building Materials Decree. In case of repair and new development of urban quay structures, the choice in materials is usually attuned to the authentic appearance of historic or old quay walls.

Materials that are often used in urban quay walls are wood, brickwork, steel and concrete. During the lifetime of the quay structure, these materials are subject to degradation.

In the following sections, degradation of wooden pile foundations and steel sheet pile structures is discussed in more detail. Additionally, it will be discussed how this degradation can be calculated in the testing and assessment of existing quay structures.

Ample research has been conducted regarding the degradation of concrete, stone and brickwork. Therefore, the degradation of these materials is not discussed in the present guideline.

2.6.2   *Degradation wooden pile foundation*

Wood is often used as a material for urban quay walls; it is applied for building foundation piles, capping beams, longitudinal beams and screens. In terms of maintenance, the degradation of wood due to chemical, mechanical and organic influences must be taken into account.

This section describes the processes behind the various forms and indicates which information must be available to make a proper estimation of the risks regarding the current and future state of the wood in the foundation.

Chemical degradation

Wood has a high resistance to chemical degradation, it is not corrosive and can withstand low pH values (>4). Wood is, however, sensitive to high pH values, but such circumstances hardly occur under water. Therefore, chemical degradation is not relevant to degradation of wooden pile foundations underneath quay walls.

Mechanical degradation

Due to specific events, a wooden pile foundation structure can be strained in such a way that stress tension cannot occur properly. When this process occurs for long enough, the structure can lose its coherence, for instance because the capping beams have been released from the piles. This can be viewed as mechanical degradation of the structure. Mechanical degradation of the wood may arise due to specific events that damage the wood, such as a crack, splintering, shearing of the wood surface or due to long-term overload, during which the wood loses strength. Jorissen [12] has created a model for this loss of strength, in which the load, wood tension and duration of the load are connected (see figure 2-14). In this model, the maximum strength is considered equal to the short duration strength. The failure line is deduced from the Madison curve and the long duration strength is equal to 0.56*short duration strength. The model shows that, when a permanent load is lower than the long duration strength, this does not lead to failure. When a load combination of permanent and changeable loads is increased for a short time (in the figure below from $t_1 - t_2$) up to above the long duration strength, but under the short duration strength, no failure or wood damage will occur. If this load is above the long duration strength for a longer period of time, damage and failure may occur on t_3.

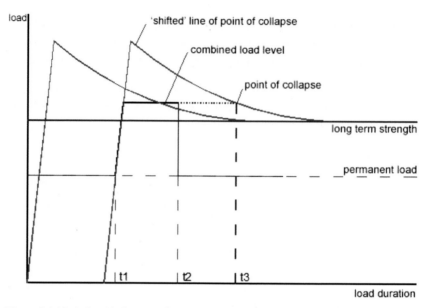

Figure 2-14 Relationship between load and wood strength (freely adapted from Jorissen [12]).

In order to make a proper estimation of the mechanical degradation of the wooden quay wall foundation structure, information on the actual construction and current state of the foundation structure is necessary. This information can only be obtained through an inspection. However, based on the applied foundation structure, a risk estimation can be made regarding the possible occurrence of damage.

Concerning the overload of the foundation wood, a risk assessment must be made in regard to the foundation structure and a reconstruction of the historical or current loads.

Biological degradation

Biological degradation is also referred to as decay and a distinction can be made between continuous and non-continuous processes.

Non-continuous decay

When foundation wood comes into contact with air as a result of a dry period, wood-degrading fungi become active, as the air contains sufficient oxygen for their metabolism. When the dry period is over and the foundation wood is once again below water, the process will immediately stop due to a lack of oxygen. The extent and duration of the dry period determine which types of fungi will occur in the wood. Infection does not play a role as these wood-degrading fungi can occur at all times and in all places. The moisture level of the wood determines the type of fungi.

42

If the dry period is such that the wood cannot dry and is thus water-saturated, this means that no air and hardly any oxygen will enter the wood. The availability of oxygen is low and the oxygen supply only comes through the air around the piles. Soft rot fungi are capable of breaking down the wood with even a limited amount of oxygen. The maximum decay speed is estimated on 10 mm/year. This speed is lower and depends upon the availability of oxygen and subsequent air-permeability of the soil and the species and type of wood.

During the dry period, there is a chance that wooden piles will dry up. In case of dehydration, air enters the wood and the oxygen in the air activates other types of fungi such as white rot and brown rot. These wood-decaying fungi are highly efficient and the speeds of decay can reach up to 100 mm/year. This value can be seen as the top limit and the actual speed of decay is therefore lower. This speed of decay is dependent upon the moisture level of the wood and the species and type of wood. Wood that has been highly deteriorated by brown, white and soft rot fungi has lost its texture. These forms of degradation are directly visible during an inspection.

However, in case of less advanced decay, visual recognition can be more difficult. The pile wood can be dehydrated via the top layer, especially around the heart, due to which the decay is not visible on the outside or measurable as a soft shell. Only with a drill sample assessment into the heart of the pile, can sufficient information be obtained about the presence and the extent of brown, white and soft rot decay. Other aspects may also be shown in the drill sample, providing more information, especially regarding the extent of a dry period. An example is the presence of blue hyphae. These fungi grow in strongly dried sapwood and can form several centimetres of new fungi tissue in warm weather.

Although blue fungi do not deteriorate wood and only live off the content substances, their appearance and distribution through the wood provides information on how far the wooden pile head has dried in the past. In order to make a proper estimation of the non-continuous decay, historic water level information must be available, including information about the type of wood and the course of the decay across the full diameter.

Continuous decay

In salt and brackish water above the sediment, pile worms and gribble may occur. Both aspects can completely destroy the piles in just a few years. Most urban quay walls are built in fresh water and these forms of decay are not relevant. However, if there is a chance of salt or brackish water being present, these forms of decay must be taken into account in the quality assessment of the foundation wood.

Continuous decay is mainly caused by bacteria, that can occur at all times and in all places. As soon as the wood is placed in the ground, these bacteria become active, just below the water surface, but also in deeper soil layers. Only in case of (minimal) water flow in the wood will the wood-deteriorating bacteria become active. Variations in the water flow lead to variation in the speed of bacterial decay. This explains the differences in decay between locations where piles have been applied between types of wood and between different wood qualities. The permeability of the wood, which is dependent upon of the type and quality of the wood, plays a role, but also the possibility of water movement in and through the soil layers. Without water flow in the wood, the speed of bacterial decay is relatively small and with increasing water flow, the speed can increase to approximately 1 mm/year. These high speeds can be found in alder and thin pine piles, but especially in screens made of permeable types of wood such as pine and beech in which a lot of water flow takes place. Although deteriorating bacteria always occur in wood and can occur anywhere, it has been shown that bacterial infection occurs more quickly in wood surrounded by soil than in wood surrounded by water. In the ground, the concentration of wood-decay bacteria is larger. In wooden piles which are completely surrounded by soil layers, pressure differences from water over the pile occur more quickly, causing water to flow inside the pile and bacteria to enter the flow. In contrast to decay in dry periods, bacterial decay in the whole length of the pile occurs. It is not yet exactly known which factors are determinative for the spread of the decay across the entire pile length, but it is clear that this spread decreases towards the tip of the pile. It is not yet possible to predict the extent to which this spread decreases.[13].

In contrast to fungal decay, wood does not lose its texture during bacterial decay. Water-saturated wooden elements may seem unaffected to the eye, but have actually lost their strength and have softened. An investigation of approximately 3300 pile heads from Amsterdam and Rotterdam, 180 of which were from water engineering structures, showed that the speed of rot caused by bacterial decay in piles underneath buildings is 0.4 mm/year on average in the case of pine wood and 0.2 mm/year in whitewood.

The speed for piles underneath water engineering structures is significantly lower than it is for buildings with an average of 0.3 mm/year for pine and 0.1 mm/year for whitewood.

Continuous decay can also be caused by soft rot in open water that is saturated with oxygen, for instance in locations with a lot of water turbulence. The speed of decay, however, is relatively low and is estimated to be 0.05 mm/year maximum and is therefore less determinative for the quality of the piles than bacterial decay.

A good estimation of the effect of continuous decay on the actual and future quality of the foundation wood can only be made based on a foundation study and wood drill sample analysis as the continuous decay is dependent upon the wood (species and type) and the potential activity on location.

Life expectancy of underwater piles with bacterial decay

Due to the construction of an urban quay wall, more variation in decay can be expected in the wooden pile foundation than is the case for foundations underneath buildings. This is due to the fact that hydrology dynamics are reflected faster in open water than in groundwater and because the oxygen level can vary much more in open water than in groundwater. Investigations on drill cores of wood can reveal the effects of these differences on the quality of the wood. Factors in these investigations are: tree age, species and type of wood and the extent

44

and type of decay across the entire wood diameter. Conclusions can be made regarding the strength gradient of the pile head as well as a quantification of the increase of the soft shell through time. The occurred decay leaves a fingerprint on the wood and the wood structure reveals blockages for decay in the wood. The extent of decay determines whether or not the blockage has been reached and whether the decay is active.

Calculations on decayed foundation wood

The remaining bearing diameter is determined based on the original diameter, the impact value and the inspection of the wood. In order to determine the wood tension, the turning point in the pile must also be known. The F30-CUR-SBR guideline [11] provides a calculation method to determine the residual diameter of the piles at the location of the turning point. The turning point is the point where the tension in the pile is highest.

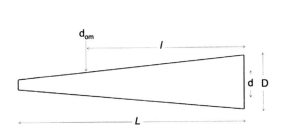 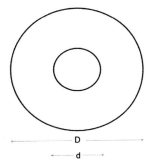

Figure 2-15 Course of decay wooden pile.

Based on the average impact values, the remaining bearing pile head surface is determined according to equations 1 and 2.

$$d = D - 2 \times (i + 5) \tag{1}$$

$$A = 0{,}25\,\pi\,d^2 \tag{2}$$

In which:

i   Average penetration depth in pile head   [mm]

D   Original pile head diameter   [mm]

d   Remaining bearing pile head diameter   [mm]

A   Remaining bearing surface pile head   [mm²]

The remaining bearing pile diameter at the location of the turning point can be determined according to equations 3 and 4. These equations make use of (in conformity with NEN 5491), [29] the original pile diameter and a tap speed of 7.5 mm/m. The location of the turning point is determined on the basis of the soil profile. Equation 3 is used for piles without bacterial decay and equation 4 when bacterial decay of the pile head has been established.

Although the course of bacterial decay across the length of the pile has not yet been sufficiently mapped [13], it is assumed that the thickness of the soft shell on the tip of the pile caused by bacteria is half of that on the pile head.

$$A = 0{,}25 \cdot \pi \cdot (d - 7{,}5 \cdot \ell)^2 \tag{3}$$

$$d_{om} = D - 7{,}5 \cdot \ell - \left(\frac{D-d}{2}\right) \cdot \left(1 + \frac{L-\ell}{L}\right) \tag{4}$$

In which:

ℓ Distance between pile head and turning point   [m]

L   Length of pile   [m]

d_{om} Remaining bearing pile diameter at turning point   [m]

For the structural assessment, the calculation value of the load, $\sigma_{c,d}$, on the one hand and the calculation value of the pressure strength of the pile parallel to the fibre, $f_{c,d}$, on the other hand are necessary input data to perform the test in accordance with equation 5.

$$\sigma_{c,d} \leq f_{c,d} \tag{5}$$

In which:

$\sigma_{c,d}$ Calculation value for the pressure tension parallel to the fibre   [N/mm²]

$f_{c,d}$ Calculation value for the pressure strength parallel to the fibre   [N/mm²]

The calculation value of the load $\sigma_{c,d}$ is determined by calculation of the combination formulas 6.10a and 6.10b indicated in the Eurocode [17]. The load factors to be calculated with are derived from the applicable parts of the Eurocode. If the value for $\sigma_{c,d}$ is derived from the calculation of the formula 6.10a given in the Eurocode with mainly permanent load, equation 5 changes into equation 5a. If the value for $\sigma_{c,d}$ is derived from the calculation of formula 6.10b given in the Eurocode with a large component of changeable load, equation 5 changes into equation 5b.

In both equation 5a and 5b, the $k_{sys} = 1,1$ (the way in which the piles work together aspect) given in Eurocode 5 in conformity with EN 1995-1-1, article 6.6, is calculated.

$$\sigma_{c,d} \leq 10,8 \text{ N/mm}^2 \tag{5a}$$

$$\sigma_{c,d} \leq 12,6 \text{ N/mm}^2 \tag{5b}$$

When calculating the load of longitudinal wood, the supporting pile surface is calculated in accordance with equation 6 with $f_{c90,d} = 4,5$ N/mm². Equation 6 may only be applied for wood which is not decayed with a thickness of >40 mm and a wood width that covers at least the entire pile head. The testing is in accordance with equation 7. Although the NEN-EN 338 [22] gives 2.5 N/mm² as the characteristic value for the pressure strength perpendicular to the wood fibre for the pine wood which is often used, with strength class C24, for instance, a higher value has been included in equation 7. The substantiation for this has been included in the CUR-SBR-F30 guideline [11]).

Because the soft shell of the pile, despite the decay, also contributes to the load transfer, a correction factor has been included in equation 6.

$$D_{longitudinal\ wood} = D - 2 \cdot (i - 10) \tag{6}$$

$$\sigma_{c90;d} \leq f_{c90;d} \tag{7}$$

47

In which:

$D_{longitudinal\ wood}$ Effective bearing diameter non-decayed foundation wood immediately on the pile

$\sigma_{c90;d}$ Calculation value for occurring pressure tension perpendicular to fiber in contact zone

$f_{c90;d}$ Calculation value pressure strength perpendicular to fiber

2.6.3 Degradation steel sheet pile walls

On the basis of the obtained inspection results, the degradation of a steel sheet pile structure can be determined and an evaluation can be made of the residual lifetime related to the structural strength and soil density.

The corrosion speed can be simplified by means of a linear extrapolation of the found values. In reality, this process is not linear and the highest values occur during the first years due to the somewhat passivating effect of the arising corrosion product. Therefore, linear extrapolation gives conservative results. Especially for relatively 'young' structures, this phenomenon must be taken into consideration.

As the starting point for the calculation of the corrosion speeds, one must determine the original average wall thickness of the component concerned. This can be done by means of pile-makeups or tables.

It is possible that another profile/wall thickness is applied than is specified in the original calculations or drawings. This must be confirmed by a first (visual) analysis of the measurement results. When the measurement results show that there are measured wall thicknesses that are thicker than originally specified, the maximum measured wall thickness must be considered as the zero thickness.

The age of the quay structure concerned is also necessary for the calculation of the corrosion speeds.

The age of the sheet pile wall is determined as of the date of dredging (the month and the year). When this date cannot be determined, the year of construction or date of construction of the structure may be used as the starting point.

Upon superimposing various 'measurement populations', for instance, recess, web and convex side of a sheet pile wall, to arrive at one general average value, the following formulas have been used.

$$n = \sum_{i=1}^{k} n_i$$

$$m = \sum_{i=1}^{k} \frac{n_i \cdot m_i}{n}$$

$$s = \sqrt{\left(\left(\sum_{i=1}^{k} \frac{2(n_i - 1) \cdot s_i}{(n-1)}\right) + \left(\sum_{i=1}^{k} \frac{2n_i(m_i - m)}{(n-1)}\right)\right)}$$

in which

n_i Size of measurement population i (e.g. recess, web, convex side)

m_i Average of measurement population i

s_i Standard deviation of measurement population i

The average rust corrosion per year is determined by the difference between the original wall thickness and the derived average (corrected) wall thickness to be divided by the age of the structure (μx / age).

Decrease in cross-section capacity

The strength of the profile is determined by the average value of the measured wall thickness reduced by an equivalent value for the influence of pitting across the surface concerned, see figure 2-16. The eventually assumed rust corrosion as a result of pitting t_{pit} has been subdivided into 9 categories. It is noted that it is assumed that this wall thickness is present across the entire surface. This is therefore a conservative approach for the calculation of the present pitting on the sheet pile structure. A disadvantage of this method is that an inaccuracy in the final corrosion speed calculated may arise due to an estimation error of the diver conducting the inspection. This inaccuracy is reinforced because two estimated values (surface intensity and depth) are multiplied for the calculation.

Depending on the influence of the steel thickness and the extent of the pitting corrosion on the stability of the sheet pile wall, it is advisable to burn or print parts for further investigation. Also see Chapter 4, section 3.5.3 where the 9 categories are distinguished.

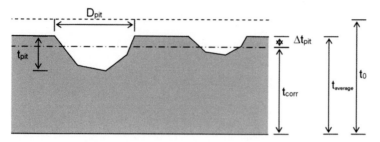

Figure 2-16 Definitions steel cross-section with pitting corrosion.

The contents of the pitting corrosion are 'smeared' across the entire surface. The contents per m² smeared thickness of the pitting is determined with:

$$\Delta t_{pit} = C * t_{pit} * A_{rel}$$

In which:

C   Factor that calculates the shape of the pitting
(= actual volume divided by the prismatic presumed volume)   [mm³ /mm³]

t_{pit} Average depth of the established pitting (estimation diver)   [mm]

A_{rel} Surface intensity (relative surface, estimation diver)   [mm² /mm²]

The occurring pitting can take on various and also capricious forms. Table 2 includes the value for C for various simplified basic forms, assuming a circular top surface of the pitting.

50

Table 2 Form factor C for pitting corrosion.

Basic shape	D / t	C	Remark
Hemisphere	2.0	0.66	Regardless the number of pits for known t and A
Half an ellipsoid	> 2.0	0.66	Regardless the number of pits for known t and A
Spherical segment	2.0	0.66	Equal to hemisphere
,,	3.0	0.57	
,,	5.0	0.53	
,,	10.0	0.51	
,,	h	0.50	= lower limit

The table shows that for the presented shapes, the shape factor C lies between 0.50 and 0.66. As a conservative assumption, it is proposed to use factor 0.66. The corrected wall thickness then becomes:

$$t_{corr} = t_{average} - \Delta t_{pit}$$

And the rust corrosion:

$$\Delta t = t_0 - t_{corr}$$

By means of the corrected wall thicknesses, t_{corr}, a statistical analysis must be made in which the corrosion speeds are calculated.

Soil density

Pitting corrosion is a highly important factor for predicting the soil density of a sheet pile structure. It is not possible to assume the average corrosion with a (smeared) addition for the pitting, as is the case for the strength calculation, because this would create a too favourable picture of the soil density.

Due to the (local) higher speed of the pitting corrosion, this is more of a guiding factor than the general corrosion.

The following aspects are important in the case of a hole developing through which soil can wash away:

- Ratio of soil behind the hole compared to the hole size (a small hole is sufficient for sand to wash away, a relatively bigger hole is needed for clay to wash away).
- Time of initiating the pitting.
- Corrosion speed of the pit.

The manner in which pitting is now shown provides an image which is too rough to make a proper predication regarding soil density. This is because (for practical reasons) a large class width was chosen for pitting depth and surface intensity. The choice for class width, the class middle and the number of classes influences the average and the standard deviation with regard to the actual situation. The error that arises as a result increases as the number of measurements decreases.

Observation:

The application of the 0.1% excess value for the measured wall thicknesses for determining the soil density was initially a starting value for the soil density (in case of missing data of pitting depths). Therefore, a very small, arbitrary excess value of 0.1% is chosen to calculate this. Later, it was chosen to predict the soil density based on the actual rust corrosion at the location of the pit. The average pit depth was determined for this purpose and it was investigated when the average wall thickness in the pit had decreased to zero with an excess of 5% above average. A check was made as to whether both methods differed from each other; this was not the case. The 0.1% excess requirement on the 'smeared' rust corrosion was virtually equal to the 5% excess requirement on the actual pit depth. This value has therefore been used in the final revision.

Table 3, from CUR 166 [7], provides an indication of the corrosion speed in various environments. These are relatively normal values. Numerous special cases are known of much higher speed of rust corrosion. Moreover, the CUR 166, CUR 221E [8], and NEN-EN 1993-5 [20] give values for rust corrosion. A wide reach must be taken into account.

Table 3 Indicative corrosion speeds in various environments (in mm per exposed side).

Intended lifetime (years)	5	25	50	75	100
Clean, fresh water around the waterline	0.15	0.55	0.90	1.15	1.40
Highly contaminated fresh water around the waterline	0.30	1.30	2.30	3.30	4.30
Salt water in temperate climate in splash zone	0.55	1.90	3.75	5.60	7.50
Salt water in temperate climate permanently under water	0.25	0.90	1.75	2.60	3.50

3 Inspection

3.1 Introduction

3.1.1 Classification inspections

Inspections are regularly carried out during the lifetime of a quay structure. These inspections can be performed or supervised by the management themselves or by an expert. For example, the management can perform a visual inspection or assessment. It is desirable that an inspection drawing is made beforehand. Possible defects can be indicated on this drawing. Additionally, it is common that a photo report is linked to this inspection drawing and included in the inspection report.

Figure 3-1 Subsidence on ground level.

Not all components of a quay structure can be easily inspected. Many of the structure's components are located underwater or underground and can only be reached by divers or by means of excavation activities. The performed inspections appear to differ greatly per municipality and demand specifications are often drawn up by an expert specifically for a project or maintenance works. Consequently, it is often the case that different inspections are carried out for the same type of structure in the same area and that the frequency of the inspections differs greatly or is dependent upon an annual budget. Additionally, other structural components may be critical per type of quay structure. Therefore, it has been decided not to link the inspection methods to types of quay structures, but to link the inspection methods to the construction materials used.

The aim of doing so is to achieve more uniformity between inspections of the same types of quay structures and between inspections in different municipalities. An important objective of this chapter is to offer guidelines for performing risk-based management and maintenance in a simpler way. This means that the manager, the testers and the inspectors must have knowledge of the objective of the inspection. The manager can use this chapter to draw up a request. Sections 3.2 up to 3.5 provide a brief description of the inspection methods per inspection class.

The inspection class to be performed must be agreed with the inspector beforehand. It is possible to perform different classes simultaneously. Depending on the objective of the inspection and the investigation, the classes are determined as shown in table 4.

Table 4: Inspection classes.

Class	Title	Objective	Frequency
0	Inspection basic geometrics	Ordering basic information	once
1a	Deformation measurement	Mapping deformations at an early stage	1x per year
1b	Visual inspection	Inspection condition and functioning	2x per year
2	Technical inspection (light)	Rough indication condition procurement	1x every 5 years
3	Full condition examination	Preparing mathematical assessment	1x every 10 years
4	Additional examination	Further examination on the basis of visible or invisible damage or result(s) from a sensitivity analysis additional examination	

3.1.2 *Inspection plan*

Prior to an inspection, it must be clear which critical components must be inspected. In doing so, attention must also be paid to the scope of an inspection. Do all wooden foundation piles need to be inspected, or does a selection of wooden foundation piles per section suffice? This aspect must be agreed upon with the tester or structural engineer and often has a statistical background. The scope of the inspection can be adjusted based on the actual inspection results. When more damage than expected is observed, the intensity of an inspection can be increased. When it becomes clear that the component concerned no longer suffices or functions properly, it is wise to stop the inspection immediately.

Client and inspector must make clear agreements beforehand. The following points must be agreed upon:
- The structures or components of the structure that must be inspected or investigated respectively.
- The class of inspection or investigation to be conducted.
- The manner of execution, especially for the measures to be taken in regard to the accessibility of the structure or components.
- The extent to which the structure is damaged (Destructive Investigation) and the manner of repair of these damages.
- An investigation plan, indicating which measurements will be taken during inspection or investigation.
- The reporting method.

As previously indicated, it is desirable to draw up an inspection drawing. The findings can be noted on this drawing. Additionally, a later photo report can be linked to this drawing. A Large Scale Base Map is often sufficient for the inspection of the quay structure.

The residual strength and stability of a quay structure cannot be determined based on a visual inspection. However, an inspection can provide input for the necessary recalculations and make it possible to assess the residual strength and residual lifetime of components of the quay structure. Additionally, the level of soil density can be determined.

The inventory and inspection class 3 are aimed at advice with regard to residual strength and the ensuing residual lifetime. A mere visual inspection of a structure of which 80% or more is located under water or underneath the bottom cannot form a basis for a definitive evaluation regarding the stability of the structure. If such an evaluation is required, this generally necessitates calculations to be carried out. Nevertheless, this does not mean that calculations must always be made. Deviations discovered during inspection that affect the safety and the functioning (regarding possible dangerous situations) must be reported by telephone to the client as soon as possible and must be confirmed immediately by fax or email. For quay structures, these could include significant washing out, scouring, excessive dry rot of the foundation wood, etc. It is up to the client to determine whether he wishes to block off the quay or take other temporary measures.

All investigations must be performed in accordance with the applicable and valid norms, guidelines or recommendations as much as possible. However, the client must realise that these were often not drawn up for quay walls and especially not for use under water.

That requires the necessary adjustments in the assignments aimed at the quay walls concerned.

Figure 3-2 Typical urban quay wall in The Hague.

3.1.3   Working with a diving team

Significant parts of the quay walls to be investigated are located under water. In most cases, the quay structure will have to be inventoried and inspected with a specialised (diving) inspection team. In order to increase the reliability of the inspection, requirements are often set for a diving team and diving team leader. Requirements may include demonstrable inspection experience and civil-technical knowledge in the field of quay structures. In general, the findings of an inspection diver must be SMART (Specific, Measurable, Assignable, Realistic, and Time-related) and the findings must be substantiated by means of measurement results, whenever possible.

In addition to professional knowledge, the diving team member must adhere to the valid legislation. For the inspection of quay walls, it is important that diving occurs with clear communication between the inspection diver and diving team leader. The inspection diver must always wear a diving helmet.

It is possible that the to be investigated quay structure is located in a used waterway. In that case, it is obligatory to use a diving flag (signal flag alfa) and to inform the waterway manager. Additionally, it is highly advisable to warn boats and other water traffic by means of signs and, if possible, to protect the diver by using a float line or boat.

3.1.4 Useful measurement equipment

During an inspection, the following can be recommended as measurement equipment:
- A plastic ruler, yellow is properly visible and does not reflect in a photo. This can be used to take most measurements. Measuring the bottom of a larger structure such as an open berth quay can be done with 2 or 3 rulers attached to a wooden slat.
- The floating capacity ensures that it remains in place on the bottom.
- In forestry, large sliding callipers are used. A plastic version up to 40 cm is suitable to measure the diameter of piles.
- Crack card for measuring crack widths, a flat stainless steel ruler for estimating the depth of the cracks. Note that the ruler can get stuck on irregularities in the crack and that the crack may actually be deeper.
- Tilting can be measured by means of an inclinometer. It is advisable to mount the meter on an aluminium ruler of 0.5 or 1.0 m. Consequently, the angle to be measured is not influenced by smaller irregularities.
- A depth gauge or measuring tape can be used for determining the bottom position. A higher accuracy is of course obtained by using a depth sounder (single or multi beam). A depth gauge or measuring tape can also be used to measure the height of the findings compared to the capping beam or waterline (take the tide into account). Please note: always convert heights to *NAP* (which will be referred to as *NAP*; the Dutch indication for this reference level).
- A prodding rod, for instance of stainless steel, 1.5 m long, ø 8 mm to determine the thickness of the mud, to assess whether the ground-retaining screen has been founded with sufficient depth, or to determine whether hollow spaces are located behind the ground-retaining screen.
- Picture and video equipment. Depending on the underwater visibility, it is possible to make recordings in order to substantiate findings in reports. In practice, at the best of times, the images have a supportive character.

3.1.5 Inspection report contents

With regard to the inspection, it is important that the inspection report provides insight into the points below:
- Overview photos or drawing.
- Front-view photo, photo layout ground level.

- Details mooring pins, capping beam, waling, anchors and bollards with ruler.
- Sketch normative cross-section in case of inspection class 2 or 3.
- Inspection drawing with damages, measurement and sample locations.
- Basic information (inspection and reporting date, inspector(s), quay number, version and report number).
- Introduction, description order and objective of the inspection.
- Findings and/or damage table.
- Measurement and analysis results, description of equipment used and calibration method.
- Possible conclusions and advice.
- Inspection drawing (as appendix to the inspection report).
- Pictures (as appendix to the inspection report).
- Measurement values and analysis (as appendix to the inspection report).
- Construction drawings (as appendix to the inspection report).

3.2 Basic geometrics of inspection (class 0)

3.2.1 In general

Throughout the years, for various reasons, a lot of information regarding existing quay structures has been lost and is no longer available. Consequently, the owner and/or manager have insufficient insight into the assets in the area. In order to determine the condition of these quay structures without usable drawings, relatively intensive and relatively expensive inspections are often required. Without the inspection results of a more intensive inspection, it is impossible to make an evaluation of the structure. It is undesirable to not conduct evaluations on structures. In Chapter 2, section 2.3.3 File management, it has already been indicated which information must be saved and gathered. For historic quay walls, of which insufficient information is available, a class 0 inspection has been included for this problem. The basic geometrics can be established with this inspection class. Moreover, the information of old structures and obstacles in the surface can be observed before drawing up a request for inspection. The problem that can occur during digging an inspection trench is that obstacles may occur in the surface, which results in additional costs for such inspections.

3.2.2 Retaining wall on spread foundations (type 1)

1) Layout and mooring facilities on ground level:

Possible differences in height, functions, and the used materials are important. The minimal inspection width, the distance transverse to the quay line, must be equal to at least twice the retaining height of the quay structure. Take all components that influence the strength of the structure into account, such as bollards, recessed bollards, mooring poles and mooring rings.

2) Measuring retaining height:

Measuring topside of the quay structure, waterline, and bottom level compared to *NAP* per 25 running metres of quay wall.

3) Course of bottom level:

Corrosion, especially caused by propeller jet load, may have occurred near quay structures. If available, a gauge map can be used for a good impression of the course of the bottom. With a depth gauge, a global indication can be obtained of the course per 25 running metres of quay wall. A survey line must be attuned to the passive ground wedge. As a rule, the same length of the ground-retaining height can be chosen for the length of a survey line (perpendicular to the quay line). If, during inspection, large scouring holes are found, immediate measures must be taken.

4) Measuring mooring facilities and quay details:

Take all components that influence the strength of the structure into account, such as bollards, recessed bollards, mooring poles and mooring rings. The maximum shipping classes for water and the quay wall concerned are connected to this. The System Breakdown Structure of the manager can be used for this.

5) Location of damages/deviations:

It is important that possible damage locations are accurately measured, such as numbered bollards, the beginning and end of the structure and bridges. Drains and trees are also often indicated on a Large Scale Base Map (digital) surface.

6) Measuring brickwork and capping beam:

The dimensions of a possible capping beam or upright course of bricks must be measured. This includes the height of the brickwork and the thickness, both at the top and at the bottom. On the land side, a test trench must be dug per 25 running metres of quay wall. Additionally, the footing course or the beam grid must be measured with regard to the *NAP*.

3.2.3 Retaining wall on pile foundations (types 2 and 3)

1) Layout and mooring facilities on ground level:

Possible differences in height, functions and the used materials are important. The minimal inspection width, the distance transverse to the quay line, must be equal to at least twice the retaining height of the quay structure. Take all components that influence the strength of the structure into account, such as bollards, recessing bollards, mooring poles and mooring rings.

2) Measuring retaining height:

Measuring topside of the quay structure, waterline, and bottom level compared to *NAP* per 25 running metres of quay wall.

3) Course of water bottom level:

Corrosion, especially caused by propeller jet load, may have occurred near quay structures. If available, a gauge map can be used to achieve a good impression of the course of the bottom. With a depth gauge, a global indication can be obtained of the course per 25 running metres of quay wall. A survey line must be attuned to the passive ground wedge. As a rule, the same length of the ground-retaining height can be chosen for the length of a survey line (perpendicular to the quay line). If large scouring holes are found during inspection, immediate measures must be taken.

4) Measuring mooring facilities and quay details:

Take all components that influence the strength of the structure into account, such as bollards, recessing bollards, mooring poles and mooring rings. The maximum shipping classes for water and the quay wall concerned are connected to these details. The System Breakdown Structure of the manager can be used for this purpose.

5) Location of damages/deviations:

It is important that possible damage locations are accurately measured, such as numbered bollards, the beginning and end of the structure and bridges. Drains and trees are also often indicated on a Large Scale Base Map (digital) surface.

6) Measuring brickwork and capping beam:

The dimensions of a possible capping beam or upright course of bricks must be measured. The height of the brickwork and the thickness both at the top and at the bottom. On the landside, a test trench must be dug per 25 running metres of quay wall. Additionally, the footing course or the beam grid must be measured with regard to the *NAP*.

7) Measuring (relieving) platform:

If a platform is present, the width of this platform can be measured. Additionally, the thickness of the floor parts, capping beams, longitudinal beam and/or water beam must be determined.

8) Measuring anchoring:

If present, the attachment, centre-to-centre distance and anchor rail must be measured. It is preferable not to dig out the anchor screens on the passive side. If this is necessary, the anchoring must be tensed again afterwards or the ground before the anchor screen must be properly compacted. Steel drawbars are currently used as an anchor. In urban areas, wooden pile yokes may have been applied, where a wooden beam functions as a tie bolt.

9) Measuring the geometrics of foundation piles:

It is important that the number of pile rows, centre-to-centre distances in longitudinal and transverse direction and the pile diameters are included. A point of attention when measuring a pile foundation is the connection of the foundation piles to the platform/weight structure.

10)   Measuring length foundation piles:

Geophysical techniques are available, where the length of a foundation pile is determined by means of a ground penetrating radar (360° GPR). Prior to this measurement, a plastic pipe is applied by means of a jet pipe or ground drill. The ground penetrating radar is entered through this pipe. When using a jet pipe, the constructor must realise that the stability of the sand layer in which the point of the pile is located will decline. The accuracy of such ground penetrating radar measurements is within several decimetres. It is advisable to assess the results together with surveys and the foundation technical possibilities in the year of construction.

The maximum horizontal difference between the ground penetrating radar and the pile to be measured amounts to 2.0 m. When determining the length of the pile foundation, possible tilting of the piles must be taken into account.

One limitation of ground penetrating radar techniques is the type of soil. In a sandy environment, the equipment generally functions well. In types of soil such as clay and peat, however, it hardly functions at all. Additionally, the ground penetrating radar can only be used in fresh water.

By means of a magnetometer, it is possible to determine the length of a steel sheet pile or other steel object in the ground (such as a reinforced concrete foundation pile).

11)   Ground-retaining screen or front wall:

The location, width and, if possible, the wall thickness of the ground-retaining screen and the manner of mounting must be included. Moreover, the foundation depth must be determined. Please also see the next section.

3.2.4   Sheet pile structure (type 4)

Steel sheet piles have often been applied in urban environments. However, this is not always visible. In some cases, the structure is finished with a concrete capping beam until below the waterline, possibly completed with a layer of brickwork.

1) Layout and mooring facilities on ground level:

Possible differences in height, function, and used materials are important. The minimal inspection width, the distance transverse to the quay line, must be equal to at least twice the active ground segment behind the quay structure. Take all components that influence the strength of the structure into account, such as bollards, recessing bollards, mooring piles, mooring rings and anchorages. The presence of anchors may increase the inspection width mentioned above.

2) Measuring retaining height:

Measuring topside of the quay structure, waterline, and bottom level compared to *NAP* per 25 running metres of quay wall.

3) Course of water bottom level:

Corrosion, especially caused by propeller jet load, may have occurred near quay structures. If available, a gauge map can be used for a good impression of the course of the bottom. With a depth gauge, a global indication can be obtained of the course per 25 running metres of quay wall. A survey line must be attuned to the passive ground wedge. As a rule, the same length of the ground-retaining height can be chosen for the length of a survey line (perpendicular to the quay line). If large scouring holes are found during inspection, then immediate measures have to be taken.

4) Measuring mooring facilities and quay details:

Take all components that influence the strength of the structure into account, such as bollards, recessing bollards, mooring poles and mooring rings. The maximum shipping classes for water and the quay wall concerned are connected to these details. The System Breakdown Structure of the manager can be used for this purpose.

5) Location of damages/deviations:

It is important that possible damage locations are accurately measured, such as numbered bollards, the beginning and end of the structure and bridges. Drains and trees are also often indicated on a Large Scale Base Map (digital) surface.

6) Determining the type of sheet pile wall:

Firstly, it is important to determine the type of sheet pile, in which the brand/manufacturer is the most important information. All brands/manufacturers have different sheet pile locks. Figure 3-3 presents a number of locks with the accompanying brand.

Subsequently, the profile must be measured. Take the working width of the pile a number of times for this purpose. Additionally, use the measurement from the convex side front to the front recess. Add the thickness of the recess to this for the height of the profile. Especially for more modern sheet piles, the type can be deduced with this method.

Points of attention for measuring a steel sheet pile structure:
- As reference, measure the thickness of the convex side, web and recess at a location where the decrease in material is minimal. This is usually just below the capping beam. If the profile cannot be determined, it is important that the angle between the web and the convex side is measured. By means of this, the resistance moment of the profile can be determined. U or Z profile, single or double piles.

- Old sheet piles are often at a right angle, modern types are more tilted (angle web and convex side). Names and types are sometimes rolled in. Take the time looking for this information, this provides certainty. Especially Larssen/Rombat, Arbed/Belval, and Hoesch used to roll in the brand and type. Different brands used the same sheet pile profiles: Larssen = Rombas France = Union Dortmond, Arbed = Belval = Columeta. Note down which type of sheet pile it concerns: U, Z, HZ, canal pile or combination wall. Cold (up to 6 mm) or hot rolled.
- Note down the height, centre-to-centre distance, angle, diameter, type and anchors (regular or grout). Convert the measures to *NAP*. Are there no anchors but instead bolt knobs in the recesses? That means that there is a waling and the anchors are invisible.

7) Determining the length of the sheet pile:

By means of a magnetometer, it is possible to determine the length of a steel sheet pile or other steel object in the bottom (such as a reinforced concrete foundation pile). For this purpose, a cone penetration test with a built-in magnetometer is used. The accuracy of the measurements amounts to approximately 5 cm.

With a length determination by means of a cone penetration test, the geotechnical characteristics of the bottom are determined. It is not always necessary to perform additional cone penetration tests.

When determining the length of sheet piles, intended variations in the length and the so-called 'stepped' sheet piles must always be taken into account. It is also possible that duos of different lengths have been placed, which is referred to as a stepped situation.

The maximum horizontal distance between the magnetometer and the sheet pile to be measured amounts to 2.0 m. For a reinforced concrete pile, this is 1.0 m.

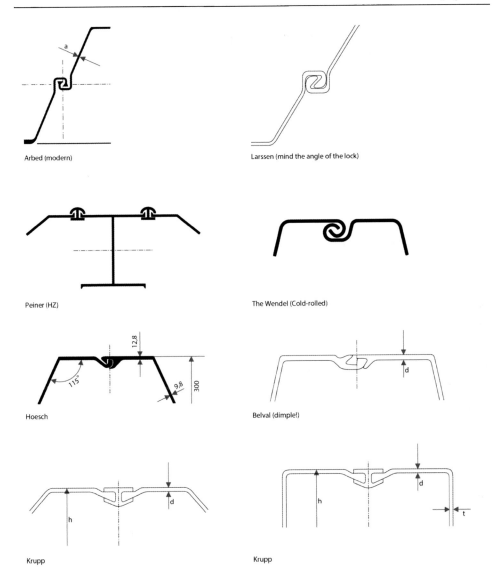

Arbed (modern)

Larssen (mind the angle of the lock)

Peiner (HZ)

The Wendel (Cold-rolled)

Hoesch

Belval (dimple!)

Krupp

Krupp

Figure 3-3 Various types of sheet pile walls.

3.3   Deformation measurements (class 1a)

The possible movement of quay walls must be periodically established by means of measurements. The results of these measurements are an important management measure for ensuring the safety and stability of urban quay walls. The causes of deformations are variable and can differ per type of quay structure. In general, we can state that instability exists and immediate measures are required if large deformations are found after years.

Possible causes are:
- extreme loads;
- construction activities;
- degradation of materials;
- scouring.

The results of the annual deformation measurements must be included in the file management.

3.4   Visual inspection (class 1b)

3.4.1   Inspection

A visual inspection aims to check which types of damage can be visibly detected and to what extent this damage occurs. The intensity of the inspection is dependent upon the age, the structure and the consequences of failure of a quay wall and the ambition level of the outdoor space on location. The quay walls must be inspected at least once a year. For new quay walls, it is possible that construction defects are discovered at the beginning of their lifetime. For old quay walls, defects resulting from degradation are often discovered. Consequently, the intensity of inspection in the beginning and end phase of the lifetime is higher than during other phases.

Upon assessing a quay structure by means of an inspection, NEN 2767 [27] can be used. The norm provides guidelines to record the extent of damage, possible defects, intensity and risk by means of a visual inspection and to give this a condition score. It is a method to arrive at a more or less uniform assessment regarding the technical state of infrastructural structure.

Please note: this is a method of which the results say something about the condition of the visible components and nothing about the stability of a structure. After all, the main part of the structure is located under water.

3.4.2   Points of attention visual inspection

During the inspection, the following aspects must be assessed:
- damages;
- deformations in horizontal and/or vertical direction;
- deformations at the location of connections, dilatations and/or cracks;
- scouring and subsidence behind the quay;
- cracks (location, pattern, shape and size);
- State of the jointing;
- Vegetation on and behind the wall (type, location, extent);
- Drainage rainwater;
- Moisture in the wall;
- Lead throughs and pipes in the wall (whether or not in use, damages to and around the lead through, material);
- Damages to stones;
- Structures attached to the quay wall (fendering, ladders, etc.) and possible damages at the attachment or base points;
- Are anchors visible and what is the state;
- State of edge finishing (end girder, top construction, upright course of bricks);
- Determining hollow space behind the front recess of brickwork and loose jointing by knocking on the brickwork;
- Taking pictures of existing situation of the wall.

These points are especially aimed at the front and top of the quay wall. If the cross-section of the wall is unknown, or if large cracks are observed in the wall, it is recommended to start a class 0 inspection.

The results of the visual inspection can be partly presented in sketches of the wall (front view with location and shape of damage, vegetation, etc.), partly by means of photos and partly with a description of the observed damage, vegetation, etc. A third party must be able to form a clear image of the quay wall based on this information.

3.5   Description of technical inspection methods (class 2 and class 3)

3.5.1   Technical inspection of brickwork

By measuring the edge joint of the brickwork, insight can be acquired into the relative subsidence of the quay structure. This can be done relatively easily by measuring the nearest edge joint around the waterline. A more accurate image can be obtained by measuring the edge joints with a liquid leveller.

When inspecting brickwork of quay structures, the common habit of renewing the brickwork from the waterline (sometimes several times) must be taken into account. This creates the risk of brickwork above the waterline looking as if it is in a relatively good state, whereas the joints under water have been washed away and brickwork may be missing locally.

Figure 3-4 Use of basalt columns in brick quay wall (Haarlem).

Erosion of the surface of the brickwork is (certainly for quay walls) often the result of frost. A relatively porous stone that becomes saturated may freeze. Partial defrosting and subsequent freezing from the outside leads to scaling of the surface.

Above the water, it is possible to test the quality of joints on location by means of a Schmidt-pendulum-hammer, see figure 3-5. Under the waterline, this piece of equipment cannot be used. However, by scratching the joints with a steel object, a comparison can be made between the situation above and below the waterline. Consequently, the extent of abrasion/washing away can be assessed.

Figure 3-5 Schmidt-pendulum-hammer.

Abrasion or washing away of the joints may be caused by frost. It must be noted that joints with relatively limy mortar wash away sooner. The binding agent dissolves and leaves behind sand. These are also referred to as sanded joints. Pay attention to the presence of shells in the mortar. Note down or sketch possible repairs or new joints. The colour of these joints is important. Cement-bound joints can be disastrous to the brickwork. The damage pattern in this case is loose joints and separated scales as a result of moisture movement through the stone.

In case of cracks forming in the brickwork, it is recommended to discover the causes before repairing the situation. In order to assess the cause of cracks, it is important that, in addition to the locations, the orientation direction, crack width, crack direction, crack depth, pattern of other cracks, erosion of edges, etc., are also noted. Moreover, cracks that are filled up again due to recrystallization must also be reported. This is in fact the self-repairing capacity of the brickwork.

It is possible to assess the quality of joints, both above and below the waterline, by conducting a qualitative chemical investigation, in which the presence of silicic acid and calcium compounds are observed. The distinction between cement and lime-bound mortar is established with this method. In order to compose a repair mortar, a qualitative analysis is also required. In this case, a mortar analysis is performed on the basis of NEN 3237 [28], where the levels of unsolvable material, silicic acid, calcium oxide and loss on calcining are established. The unsolvable material is sifted, due to which the grain gradation of the sediment material can also be established. The composition of the repair mortar can be calculated from the collected information.

For objects with a higher historical value, determining the share of the unburnt lime/binding agent and sand is important. A finely ground preparation can be made of the mortar which can be petrographically inspected. In such an investigation, the structure of the material is mapped by means of a polarisation microscope and the mineralogical composition of the mortar components is determined.

In order to determine the pressure strength of brickwork in a wall, a core of ø 200 mm is required. An arrow must indicate the top at the front of the core. In the laboratory, a core/sample of ø 100 mm is drilled in vertical direction from a horizontal core of ø 200 mm, see figure 3-6. This core is finally sawed to length, flattened and pressed.

Figure 3-6 Pressure strength brickwork, vertical cores from horizontal core ø 200.

For specific information concerning the stability of stone structures, please refer to the CUR Recommendation 73 [5].

3.5.2 Technical inspection of wooden foundation piles

It is once again emphasised that this handbook has been composed in such a way that it matches the F30-CUR-SBR guideline as much as possible for the inspection of wooden pile foundations underneath buildings [11]. Due to the differences between both foundation types, this guideline cannot be completely followed. The large number of applied piles that are partly in open water and in two zones (before and behind screen) are specific to urban quay walls.

Furthermore, inspections are conducted under water due to which the standard impact hammer cannot simply be used.

Within the specific approach of the assessment of foundations underneath urban quay walls, the sample size is an important aspect. The residual diameter for a pile population is determined and the future loss is predicted based on the penetration values and wood analyses on drill cores. For determining the sample size, the following has been taken as a starting point.

Sample size

Research shows that a relatively greater variety of wooden piles has been applied underneath quay walls compared to the variation of wooden piles underneath buildings. This is probably due to the various deliveries needed to acquire such a large number of piles. Moreover, research shows that [14] especially whitewood and redwood are found, but wood types such as pinewood, Douglas fir, oak and alder also occur and even sporadically maple, elm and ash are found. Although piles with larger diameters (20-35 cm) are often found underneath quay walls, smaller diameters are also found. Klaassen [14] demonstrates that the tree age of over half of the piles is < 40 and is > 80 in less than 10% of the cases. The pile quality is currently stipulated by NEN 5491 [29], but it is clear that such a quality was not always applied in the past. Whether piles were delivered directly from storage can no longer be determined, but this is crucial for predicting the sensitivity to bacterial decay, as storage above water has a (dehydration) effect on the permeability of the wood.

The circumstances for wood decay are different when the wood is found in front of and behind the water-retaining screens. The state of the piles at the waterside may deviate from the state behind the screen.

Many quay walls are older than 100 years. When the historic urban quay walls were built, transport by road was difficult due to the absence of lorries and a proper infrastructure. Most of the wood was transported by water, from Europe on rafts and from Scandinavia on ships. Although little information is available on the storage of wooden foundation piles in that time, we assume that all the wood was rafted and stored underwater. Only the Northern European wood spent limited time above water during transport from Scandinavia to the Netherlands.

Sample size foundation pile samples

The thickness of the decayed, non-bearing shell is determined by means of a random sample of wood samples from foundations piles where the characteristic value approach has a 5% higher and lower limit in conformity with table 2c of Eurocode 7 [32] per wood type population. Based on the CUR-publication 236 (section 3.4.1, [10]), a sample size of 3% is assumed (with a minimum of 3 pieces) and assuming 2 rows of piles that are actually present or accessible.

Practical example

For a quay wall with pile row distances of approximately 1 metre, this means that 2 wood samples per 15 running meters are included, of which half is initially tested. If the analysis shows that multiple wood types or diversity in decay is present, the second half of the wood samples is then also analysed. The minimum number of wood samples to be analysed per structure is 3.

The visible parts of a wooden foundation pile, or the parts that have been excavated, can be inspected relatively easily. It is important to ascertain whether the top of the wooden foundation pile shows visible decay. The direction of the damage is usually oriented from the outside towards the inside. The edges of the pile head have crumbled or have become eroded. When the groundwater level or the level of the harbour water has decreased, compared to the water level during the realisation, the part of the wooden pile foundation above water is also often affected.

Deviations that may influence the strength of the structure must be noted. Think of broken or cracked piles, piles that are not or only partly underneath the timber beams, space between the piles and the timber beam (negative friction), etc. There is also a chance that the piles have been decayed by gribble or pile worms. This exclusively occurs in a salty or brackish environment.

A suitable location for performing penetration measurements is approximately 10-15 cm underneath the head of a wooden foundation pile. The penetration measurements must be divided across a surface the size of a hand. The advisable number of measurements amounts to approximately 6 to 10 measurements per pile. The penetration meter, see figure 3-7, must be suitable for activities under water. For measurements with both a Pilodyn as the Specht, deviations must be taken into account. If the Pilodyn is used, no information is available regarding the extent of the deviations underwater. If the Specht is used, it is known that the deviation amounts to 40%. Given the equal functioning of both pieces of equipment, it is likely that the Pilodyn (which is no longer produced) also has a significant deviation. Due to this deviation, the measurements are less reliable and can only be used to confirm wood sample analyses. It can be said that it is relatively difficult to measure whether piles display significant decay. As the thickness of the soft shell (the decayed wood) increases, the equipment has less capacity to properly and accurately depict the decay. Currently, an increase of the deviation must be taken into account when the condition of piles decreases.

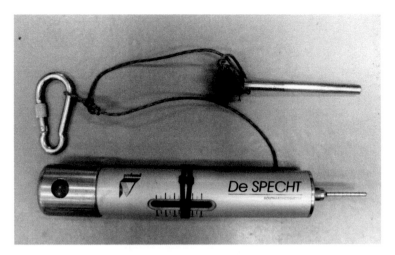

Figure 3-7 Penetration meter

Taking and analysing wood samples

The initial inspection costs for a foundation inspection of a quay wall are relatively high. This means that it is advisable to take sufficient wood samples immediately. Taking additional wood samples at a later stage brings unnecessary costs.

The wooden piles on the waterside are generally much more accessible than the piles on the landside. Additionally, as several studies have shown, the piles on the waterside are more decayed than the piles on the landside of the structure, especially when the pile heads are free in the water. In case of anaerobe bacteria, this can theoretically also be reversed.

The location in which the wood samples are taken must be attuned to the geometrics of the quay structure and is dependent upon the available space. If the connection between the pile and the capping beam is realised by means of a spring bolt, 20-25 cm below the pile head is the minimum distance in order not to damage the drill. In case of a pin-hole connection, a concrete platform or concrete component, a distance of 10-15 cm below the pile head is recommended.

It is important that the wood sample is drilled perpendicular to the foundation pile, through the heart of the pile. In doing so, the sample length is at least half of the diameter of the pile. Consequently, the amount of core and sapwood of the trunk can be measured.

The drill which is most used is a increment borer ø 10/12 mm, see figure 3-8. For drilling, a sharp drill must be used to prevent the sample being pulverised or breaking into many pieces. When the foundation piles are too hard (for instance oak wood), the samples can also be taken by means of a plug cutter driven by an air or hydraulic drilling machine.

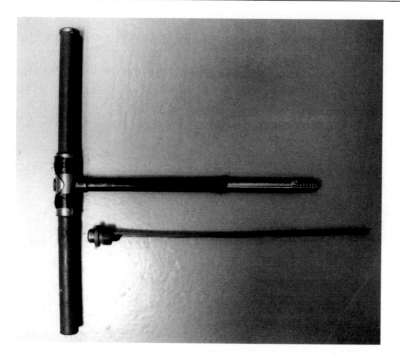

Figure 3-8 Increment borer

The samples must be stored in a tube submerged in water from the environment. The pile diameter of the tube must be chosen such that possible loose parts do not touch. The wood samples must be stored in cooling.

Due to the small diameter of the drill hole, the research method cannot be considered to be destructive. However, it must be noted that sporadically smaller pile diameters are found, such as ø 150 mm, where this observation does not apply.

It is advised to take 2 wood samples for every 15 running metres of quay wall, one of which must be tested. If the analysis shows the presence of multiple types of wood or diversity in decay, all other samples must also be analysed. The minimum number of wood samples to be analysed per structure is 3. For long quay walls (> 150 m), the number of samples can be reduced by 25%. Especially for data collection for recalculation, the relatively small sample size must be taken into account. This is based on the numbers as described before under the header 'sample'.

Inspection of timber beams

Penetration measurements on timber beams (both longitudinal beams and capping beams) must be performed in the middle of the beam above the pile as much as possible. A point of attention is that the no measurements have to be conducted on end grain wood. The remaining timber beams are important to the mutual cohesion but not for the bearing capacity of the structure. Due to the arching of the brickwork, a relatively small part of the load is transferred to the piles by the part of the capping beams not located above the piles. The minimum number of penetration measurements is 6.

Figure 3-9 Horizontal movement in a quay wall, cause of problem being capping beams or a horizontal load which is too heavy.

The visual inspection of the timber beams has a number of points for attention. For example, it is important to pay attention to the nature of the connection between the pile and the capping beam. The damages which occur are described below.

Deterioration of capping beam heads

The heads of the capping beams are often deteriorated. How much these protrude from the pile does not necessarily influence the stability. There is a danger of deeper pontoons or ship hulls, that run straight down, sailing into the capping beam heads. Consequently, movement, fracture or accelerated decay of the heads may arise. If the capping beam heads show such decay that this influences the connection between the pile and the capping beam, it is useful to estimate the percentage of the decrease in load-bearing surface. In that case, all connections must be inspected.

Significant decay piles and timber beams

If piles and the timber beams above display significant decay, it is possible that the pile is crushing or perforating the capping beam due to the increased concentrated load.

Pin-hole connection

In case of a pin-hole connection between foundation pile and a capping beam, a doubling of the areas that can be affected in combination with 1/3 less material of the capping beam must be taken into account.

Construction defects bearing

A longitudinal beam usually only rests on part of a batter pile. A corner has been sawed from the batter pile on which the longitudinal beam rests. Relatively many construction defects are found in this part. Due to the relatively smaller surface on which the longitudinal beam rests, this must be taken into account upon the assessment of dry rot in the pile. Also pay attention to the quality of the connection between the batter pile and capping beam, this is usually an end thread.

Positioning wooden piles

Piles that are only partly underneath a capping beam. It is important to discover the cause. Possible causes are a construction defect, sliding, movement of the pile due to a mechanical cause or dry rot.

Pile break

Especially in the front row of piles and/or batter piles that stand free in the water, the deeper pontoons or hulls of ships that go straight down may sail into the piles causing them to break. If such damage is established, a 100% inspection is required, given the negative influence several consecutive broken batter piles have on the structure.

Corroded ring

The corroded ring around the pile head has been applied during construction to prevent the pile head from splitting during striking of the spring or tap bolt. Therefore, it is presumed that the corrosion of the ring has no consequences for the lifetime of the structure.

Connection batter pile-longitudinal beam

The head of a batter pile has usually been sawed in and connected to a longitudinal beam by means of a stud bolt. This connection is important to the contribution the batter pile makes to the stability of the structure and, for this reason, must be closely inspected (cleaned).

Inspection of wooden ground-retaining screen

The ground-retaining screen of historic urban quay walls in most cases consists of wooden sheet piles, connected by means of a tongue and groove system. The wooden screen is attached to the piles, the longitudinal beams or the capping beam at the topside. Other occurring structures are horizontal piles or concrete sheet piles. After World War II, many steel sheet pile walls were also applied as ground-retaining screens.

During the inspection, the condition of the piles and the quality of the connections themselves and the connection to the topside must also be inspected. Moreover, the piles on the bottom must be sufficiently clamped. The number of piles to be inspected is dependent upon the accessibility. In case of an overbuilt slope, it is sometimes possible to access the ground-retaining screen from the waterside. If this is not the case, the extent must be coordinated with the number of inspection slits.

The quality of the wooden sheet piles is determined by means of penetration measurements related to the thickness of the piles. If space has arisen due to dry rot between the piles or at the top of the piles, then the screen will no longer be sufficient. The ground-retaining screen is no longer geometrically closed. It is important to inspect the remaining wood to determine whether this offers sufficient quality for the possible sustainable repair of the wooden ground-retaining screen.

If space has arisen at the top of the ground-retaining screen, or cracks between the sheet piles, this can lead to the washing away of soil and hollow spaces behind the sheet piles can occur. Hollow spaces behind the ground-retaining screen are measurable by means of a ruler gauge or in larger cases by means of a bradawl. It is important that the inspection diver records which types of soil are found in order to assess the expected speed of washing away.

Larger hollow spaces can result in subsidence on surface level of several cubic metres. The above street can conceal a hollow space due to mutual coherence or the road foundation for a long time. Consequently, it is possible that a large hole suddenly arises behind the quay structure and abruptly caves in, resulting in a safety risk. If cracks have been established in the ground-retaining screen at the inspection locations, it is advisable to expand the inspection of the ground-retaining screen to a 100% inspection. The result of this inspection component can be a drawing on which the size of the found hollow spaces is indicated in the form of a spot map. Additionally, the pieces of sheet pile wall in which cracks have been found are indicated. With a geo textile, these holes can be sealed (temporarily).

Due to for instance dredging activities or erosion due to screw radiuses, it is possible that the bottom level is (locally) lowered. Due to sedimentation, the bottom can be locally heightened. Local lowering is a point of attention for the inspection diver. The lowered bottom directly influences the stability of the ground-retaining screen. It is noted that a (local) lowering of the bottom level has significant influence on the stability of the entire quay structure.

3.5.3   Inspection steel sheet pile

Upon the assessment of a steel sheet pile, skewing arisen in the construction phase must be taken into account. The sheet piles coming loose from the lock form a risk. Wooden fendering, especially in vertical direction, that has been attached to the sheet pile with welded strips display regular cracks in addition to the welds of the strips. These have arisen due to fatigue as a result of the berthing forces. Eventually, this also leads to holes in the sheet pile resulting in washing away of the soil behind the pile.

Figure 3-10 Measuring uniform corrosion around the waterline.

In order to form a proper evaluation of the occurring corrosion of steel sheet piles and with that the state of the structure, sufficient measurements must be conducted for statistical processing. For this purpose, the quay structure is divided into a number of measurement trenches to be defined beforehand (spread across the length of the quay) and standard measurement levels (across the height of the quay). The measurement trenches are chosen in such a way that these are located next to an object recognisable for the diver (such as bollards, king piles, stairs or recessed bollards).

Due to this method, a grid/matrix arises across the front of the quay wall. The intersections of the vertical trench lines and the horizontal level lines form the measurement points on which the measurements are conducted.

For every quay configuration, an estimation must be made of the number of trenches and measurement levels to be chosen beforehand. In practice, it is possible to compare the representative results in case of trench distances of approximately 25 metres (length of fields) and vertical distances in the depth of 2 to 2.5 metres. A distinction can be made between the following measurements:

- General corrosion measurement
- Position measurement

Uniform corrosion

Uniform corrosion, the equal rusting of steel, can be measured with ultrasound technique (see figure 3-10). When a thick layer of corrosion products is present, sometimes in combination with algae and dirt, incorrect values can occur. The corrosion product is usually thicker, approximately 6 to 7 times the rusted thickness.

The production processes in the early years were less controlled, meaning that production errors often occurred. For example, due to mistakes in the rolling process, delaminations arose, so-called doublings. Due to the presence of doublings in the piles, for instance, it is possible that values are measured with a high level of variation. In such cases, it is essential that measurements are once again conducted left, right, above and underneath at approximately 1 metre.

When interpreting the wall thickness measurements of the steel sheet pile walls, deviations in thicknesses must be taken into account. It is possible that a certain profile with an under or over thickness was ordered. Moreover, it is possible that multiple types of sheet pile wall have been used.

In order to obtain a correct view of the original thickness of the sheet pile wall to be measured, it is advisable to conduct a number of reference measurements at a location where it is likely that the wall thickness has decreased.

Pitting corrosion

Pitting corrosion is also referred to as pitting or cavitation. In case of pitting corrosion with a depth of over 1 mm, it is not always easy to conduct the wall thickness measurements. For this reason, it is important to be alert. The (diving) inspector has a tendency to look for a better section of the sheet pile wall to obtain measurement values. Thickness measurements performed on a better part of the sheet pile present a view of the pile thickness which is too optimistic. This is an undesirable situation.

Therefore, it is advisable to inspect the weakest sections of the sheet pile structures with significant/poorly measurable pitting corrosion by burning out parts of the structure. The rule is: less than 50% measurable matter gives a too optimistic view of the sheet pile wall.

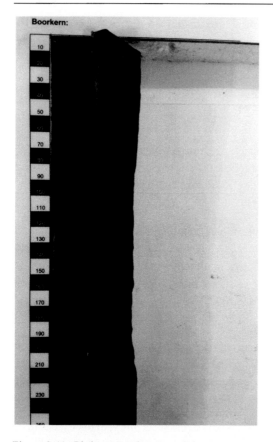

Figure 3-11a Pitting corrosion around the waterline.

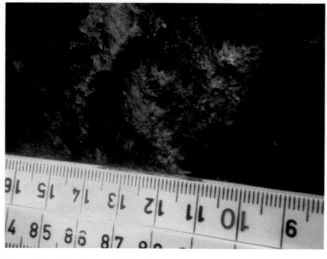

Figure 3-11b Surface with pitting corrosion.

Causes of corrosion

Significant corrosion near the transitions from convex side to web has been found in various sheet pile structures. There are known cases in which 12 mm thick sheet piles have decayed due to rusting in just a few decades. Holes in the piles have arisen near the transitions between the sheet piles. The cause of this degradation is most likely tension corrosion. This tension could have arisen due to a temperature which was too low during the rolling process.

In the sheet pile locks, the corrosion is usually more advanced than at the surface. The cause is the moist environment. This is also referred to as crevice corrosion.

In certain cases, the holes for the anchor poles are too large. This may cause a flow of ground-water, resulting in the accelerated degradation of the sheet pile due to the moist environment.

Other causes of corrosion may be impurities (osmoses) in the material, failing conservation and local salt concentrations.

General measurements

On every measurement trench and accompanying measurement level, the diver must conduct 3 measurements per section of the quay structure. Furthermore, a measurement must be conducted 50 mm above and 50 mm below the measurement level, for every measurement trench and accompanying measurement level.

This is done to compensate for a possible erroneous measurement, for instance caused by a locally situated crater, at the location of a measurement position. The 3 local measurements per point are intended to determine the local wall thickness as exactly as possible and all 3 points are included in the statistical processing. If a point is immeasurable, due to a local deep crater, the remaining points that are measurable nearby are representative for the local wall thickness.

For steel sheet piles, the wall thickness is often measured at the location of:
- The recess of the sheet pile, this is the side that is located mostly in the groundmass as seen from the water.
- The convex side of the sheet pile, this is the side that sticks into the water most.
- The web of the sheet pile, the side that connects a recess to the convex side.

It is desirable to record these locations accurately to ensure that repeated measurements can be conducted in the future. It is also possible to use moulds to measure the exact locations again. Figure 3-12 includes an example of such a mould.

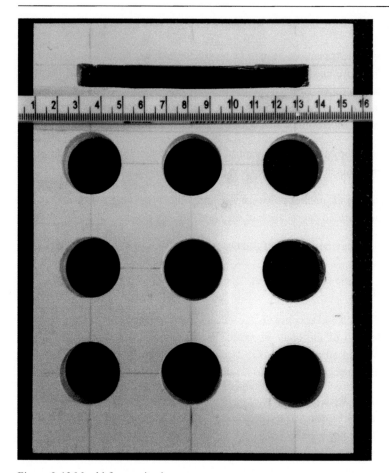

Figure 3-12 Mould for monitoring.

In practice, it has been shown that systematic error may also occur in formulated measurement methods. This is caused due to the taster with which the measurements are conducted being unable to conduct a measurement at the location of pitting. The diver will have the tendency to move the taster until it once again registers a value and therefore arrives at too optimistic values.

Initially, these values are superimposed on the values obtained with the measurements. The extent of the pitting must be based on an evaluation of the diver, who makes an estimation of depth and surface intensity of the established pitting. A division has been made concerning the surface intensity and depth, as indicated in tables 5 and 6.

Table 5 Class division intensity

Class	Intensity pitting
i1	Up to 30%
i2	> 30% and < 70%
i3	> 70%

Table 6 Class division depth

Class	Depth pitting
d1	< 2 mm
d2	> 2 mm and < 4 mm
d3	> 4 mm

Consequently, the rusting which is eventually assumed as a result of pitting Δt_{pit} belongs to one of the 9 categories, see table 7. In order to determine the cross-section reduction resulting from corrosion based on the estimated and measured values, please refer to section 2.6.3.

Table 7 Division prismatic assumed pitting Δt_{pit}.

Matrix Δt_{pit}	i1 (0,30)	i2 (0,50)	i3 (0,85)
d1 (1 mm)	0.30 mm	0.50 mm	0.85 mm
d2 (3 mm)	0.90 mm	1.50 mm	2.55 mm
d3 (5 mm)	1.50 mm	2.50 mm	4.25 mm

Position measurements

In order to compensate for the disadvantages of the described method and to obtain a more refined view of the occurring corrosion, additional measurements with enlarged surface density have been performed in locations determined beforehand. Thanks to the statistical processing of this data, an image of the locally present corrosion may be obtained.

In general, it is attempted to perform 1 to 3 position measurements per measurement level. The measurement area is the same for every measurement level. Measurements must also be conducted 50 mm above and 50 mm below the measurement level. As is the case for the chosen trenches for the general measurements, the locations for the position measurements have also been determined in a random manner.

The number of wall thickness measurements to be conducted is at least 9 per location. This number is based on the diversity that one may encounter during measuring. If the spread of the numbers is for instance, ± 15% from the median, it is advisable to increase the measurements by up to 25 per measurement location.

For the measurement location, a surface the size of a hand is advised. It is desirable that these locations are accurately recorded, allowing repeated measurements to be conducted in the future.

In order to acquire a complete view of the decline of the sheet pile, usually the minimal, average and maximum value, it is important to indicate the decline in absolute value and percentages. See the example in table 8.

Conservation

The condition of the conservation can be determined by means of NEN-EN-ISO 4628-2 [23], NEN-EN-ISO 4628-4 [25] and NEN-EN-ISO 4628-5 [26] concerning the evaluation of blister forming, crack forming and flaking. The total surface area is important, where it must be noted whether the damage concerned is spreading or occurs locally. A black conservation type may contain tar. This can easily be determined by scratching off a layer and smelling the flakes. Components with tar can be easily established with this method. If the conservation type contains tar, the nature can be established by wiping the surface with a cloth with acetone or thinner. When the conservation layer dissolves, the conservation on coal tar basis is different than tar epoxy. Knowledge of this is important in choosing the system for repair.

3.5.4 Inspection of concrete

In general

In order to estimate the cause of crack formation in brickwork, it is important that the orientation direction, crack width, crack direction and crack depth, pattern with other cracks, erosion of the edges, etc., are noted in addition to the locations. Cracks that have filled up due to recrystallization must also be noted. This is the self-repairing capacity of the concrete. The allowable crack width is related to the type of reinforcement and the environmental class of the concrete structure.

In order to make an assessment of the lifetime of the concrete component, it is important to determine the concrete cover. If there is a reason to do so, the carbonation depth can be determined. In a salt environment (by the sea, thawing salts), the chloride penetration is relevant. In both cases, a relatively low coverage or possible damages could be the cause.

Table 8 Example table of measurement values.

Lowest measurement value [mm]		highest measurement value [mm]		Average measurement value [mm]		Initial thickness [mm]		Decline [mm]		Decline [mm]	
Convex/ recess	Web	Convex/ recess	Web	Convex/ recess	Web	Convex/ recess	Web	Convex/ recess	Web	Convex/ recess	Web

An additional inspection with regard to the residual lifetime determines the permeability by conducting water penetration measurements.

For the mathematical inspection of the concrete structure, the configuration and strength of the reinforcement is required. Additionally, it is desirable that the pressure strength of the concrete is known, and if necessary also the tensile strength. These tests are generally destructive, see figure 3-13.

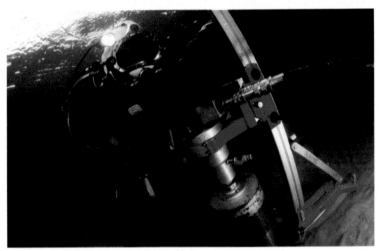

Figure 3-13 Destructive test under water.

Determining reinforcement configuration

Above the water, the location can be determined by means of field strength metres, a ferroscan or a concrete radar. For situations underwater, there is no measurement equipment available and the location of the reinforcement can be determined by means of chiselling out a piece of the concrete to be tested. It is noted that underwater repair is relatively intensive and often does not provide a satisfactory result.

Quality reinforcement

Both above and below water, the reinforcement can be ground or chiselled out. The test samples of 30 to 40 cm long must be undamaged. By means of a tensile test, the mechanical properties of the material can be determined. The extent of the corrosion must be visually determined and it must be noted whether it concerns smooth or profiled reinforced steel.

Compression and tensile strength of concrete

The compression strength, axial tensile strength or splitting strength is determined from a core of ø 100 mm. For information regarding the determination of the compression strength of concrete, please refer to the NEN 5968 [30]. By means of air or hydraulic (core) drilling equipment, it is possible to drill cores underwater.

In order to assess the residual lifetime, it is advisable to draw up a description of the cores, an orienting petrographic test.

If, during the visual inspection, damages are found that point to Alkali-Silica-Reaction (ASR) or ettringite in the concrete, extensive petrographic testing may then be required. As is the case for mortar, a thin sample is assessed under a polarisation microscope.

For more information concerning inspection and testing of concrete structures, please refer to CUR-Recommendation 72 [4].

4 Testing and consideration framework

4.1 Introduction

In the previous chapter, four different levels or classes of inspections were named. If it appears that sufficient information is available from an inspection (or possibly from the inventory) to evaluate the structure, the structure can subsequently be tested. In this chapter, the testing method will be described in which the failure mechanisms per type of quay wall are considered. The norms and criteria for testing are discussed, established on the basis of the state of the structure.

4.2 Loads in urban areas

4.2.1 Introduction

Many quay structures in urban areas were designed for facilitating shipping or as a protective bank in an urban water system. Shipping has now often been moved to locations outside the city. Consequently, the load due to shipping, storage and transhipment has decreased, whereas the load due to traffic has increased. Additionally, the traffic load and intensity have increased significantly during the past years. The loads on a quay structure mainly consist of the forces exercised by the ground, groundwater and harbour water, as well as ice and anchor forces on the quay structure. Exceptional load cases, such as subsidence resulting from extra dredging or flow, can result in the risk of instability. Please see practical example The Hague in section 5.6.1.

In addition to these permanent loads, the loads can also have a temporary character, such as vertically oriented forces on the cap of the quay structure, hawser forces on bollards, mooring forces, tree loads (wind) and traffic loads. In order to evaluate the stability of existing quay structures, it is important to map the current load. Many loads are described in the general guideline for sheet pile structures (please refer to section 3.2 of the CUR-publication 166, part 2, as well as CUR-publication 211 chapter 6).

The main loads described in these guidelines are:
- Load due to soil.
- Surcharge and traffic load.
- Differences in water level of harbour water and groundwater.
- Tree load.
- Ice load.

- Mooring load.
- Hawser load.
- Propeller jet load.
- Loads due to earthquakes and vibrations.
- Loads due to condensing.
- Loads due to swelling ground.
- Loads due to temperature changes.

However, in urban areas, specific loads may occur that are not described in, or that deviate from, these general guidelines. In this chapter, the specific loads that may occur during testing and repair of urban quay walls are discussed.

4.2.2 *Water level differences between harbour water and groundwater*

Quay structures in urban areas often border on an inland harbour or waterway, which is hereafter referred to as harbour water. When choosing calculation values for differences in water pressure across the quay structure, long-term observations of the groundwater levels in the immediate vicinity of the structure must be taken into consideration, also when a drainage system is constructed. For calculating the loads due to water level differences between harbour water and groundwater, the following parameters are needed:
- The rising height(s) in the various ground layers on both sides of the quay structure.
- The level of the harbour water.
- The volume weight of the water.

Additionally, water overpressure or water underpressure must be taken into account, which is dependent upon changing water levels, the supply of groundwater and the permeability of the ground. Water over-tension may occur in layers that are poorly permeable or are bordered by poorly permeable ground layers. Water over-tension can occur in the consolidation phase after the application of a surcharge or by the local softening of the ground during pile driving. Due to excavation or pumping, the water tension may decrease. It is also observed that water tensions in a structure calculation must be levelled out.

Design values

The design value of the groundwater level is not allowed to exceed the lowest occurring water level and, in tidal areas, may not exceed the multi-year average of the lowest low water spring of every month, or LLWS.

The design value of the harbour water level behind the quay structure and the minimal pressure difference across the quay structure are dependent upon the water permeability of the soil and the possible presence of a drainage system.

Low permeability without drainage

For quay structures that retain ground with a low permeability, such as silt and clay, it must be assumed that the groundwater level behind the quay may rise to ground level. If it can be demonstrated that this maximum groundwater level does not occur, for instance by means of long-term measurements of the rise level in these layers, this assumption may be deviated from.

Low permeability with drainage

If it has been assumed in the design that a drainage system has been constructed which continuously keeps the groundwater pressure low, (maintenance) provisions must be arranged for in order to assure an effective functioning of this drainage system during the lifetime of the structure. The consequences of a possible blockage or failure of the drainage system must be taken into consideration during the design/testing. An analysis must always be made in case of a poorly functioning drainage system, which is considered as a special load.

High permeability

For quay structures that retain ground with a high permeability, a groundwater flow analysis must be made in order to determine the occurring differences in water pressure. If a drainage system is applied, this must be included in the groundwater flow calculations. In tidal areas, a relatively high groundwater level in combination with a fast declining harbour water level must be taken into consideration. If no groundwater flow calculations have been made, it must be assumed that the groundwater level behind the wall may rise to ground level, unless a reliable drainage system has been constructed at a lower level.

Poorly functioning drainage system due to quay structure

During the design lifetime of the drainage system, a possible blockage or failure of the drainage system must be calculated. In this calculation, reduced safety factors may be included by considering them as special load. In the case of a poorly functioning drainage system, the groundwater level design must be considered as being equal to the situation without a drainage system. If reduced safety factors are applied after the lifetime of the design, it must be demonstrated that the drainage system still functions properly.

In that case, it must be demonstrated that sufficient (maintenance) provisions are made to safeguard the effective functioning of the drainage system after the lifetime of the design has been exceeded.

Properly functioning drainage system due to quay structure

The harbour water level is with the presence of a drainage system of course no different than it would be without the presence of a drainage system. If a properly functioning drainage system is present due to the quay structure, the groundwater level design must not be lower than the level of the bottom of the drainage system + 0.3 m.

- With a properly functioning drainage system, a calculation value of at least 5 kPa must be used for the pressure difference when small differences in the harbour water level occur (average level harbour water – lowest occurring harbour water level < 1.0 m).
- With a properly functioning drainage system, at least 10 kPa must be used for the pressure difference when relatively large differences in the harbour water level occur (average level harbour water – lowest occurring harbour water level > 0.1 m).

4.2.3 Traffic load

Under normal usage, traffic load on quay structures must be considered as a changeable load. Loads caused by road or tram traffic can transfer vertical, horizontal, static and dynamic forces onto the quay structure.

According to current legislation and regulations in the Netherlands, the maximum allowable mass of a loaded vehicle which may use the public road without restrictions is 50 tonnes. If the total load exceeds 50 tonnes, it is legally obligated to report the axle loads, axle distances and axle type to the appropriate authority. In order to determine the load caused by road traffic on urban quay walls, the design method of abutments of concrete bridges is matched, which is similar to that of a quay structure in terms of geometrics. This design method matches the traffic load as prescribed in Eurocode 1 [18]. If such a heavy load is physically impossible on a quay wall, a different approach can be chosen in which the actual occurring maximum load is calculated.

The load caused by road traffic differs per location based on:
- the distance between the carriageway and the quay structure;
- the composition of traffic, such as the percentage of lorries and construction traffic;
- the traffic intensity;
- the number of lanes;
- the circumstances, such as speed and queues;
- the expected maximum weights of the vehicles;
- the influences of signage;
- the influence of traffic lights.

By means of the characteristic value of the vertical load, it can be tested whether the existing quay structure meets the desired functionality. Four load models are distinguished:
1. Concentrated loads and equally divided loads, the majority of which consists of lorry and car traffic. This concerns global and local effects.
2. A single axle load with a certain contact surface of the tyres to consider the dynamic influences of ordinary road traffic.
3. Complete axle loads that model the special vehicles that follow a route on which exceptional loads are allowed. This model must only be used if the client has specifically

requested it. In case of doubt as to the applicability of this load model, this must be discussed with the client/manager.

4. Load caused by a crowd. This model must only be applied if the client has specifically requested it. In case of doubt as to the applicability of this load model, this must be discussed with the client/manager.

Load models 1 and 2 are discussed in further detail below. For more information on load models 3 and 4, please refer to the Eurocode 1 [18] concerning traffic loads on bridges.

Load model 1

The loads from vehicles are determined by means of a chart of the wheel axle system and the wheel loads. Per lane, a so-called tandem system must be used for calculations, see figure 4-1.

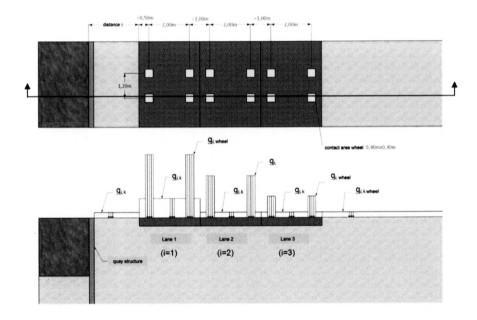

Figure 4-1 Schematic presentation of the axle system in conformity with load model 1.

This tandem system must be placed in the least favourable position in the traffic lane. In the subsequent discussion, it is assumed that every tandem system has an identical wheel configuration. The contact surface of each wheel must be considered as a square with sides of 0.40 m. When the area of influence of the axle loads cross each other, these must be superimposed.

The following applies to every axle load:

$$q_d = \gamma_Q \cdot \alpha_q \cdot q_k$$

In which:

q_d Calculation value changeable load

γ_Q Partial safety factor changeable load

α_q Correction factor for the number of lorries per traffic lane per year

q_k Characteristic value changeable load

Table 9 can be used for the determination of the characteristic values for the line loads resulting from wheel pressure and the line loads resulting from equally divided traffic load. Usually, the above configuration is the guideline. However, in case of a relatively long relieving platform, various load combinations of traffic lanes must be used for calculations.

Table 9 Loads prescribed axle system load model 1.

Position	Axle load tandem system	Wheel load tandem system	Wheel pressure per surface b*l (b=0,4, l=0,4)	Equally divided load
	$Q_{i;k}$ [kN]	$F_{i;wheel;k}$ [kN]	$q_{i;wheel;k}$ [kN/m²]	$q_{i;k}$ [kN/m²]
Theoretical lane number 1 (i=1)	300	150	940	9.0
Theoretical lane number 2 (i=2)	200	100	500	2.5
Theoretical lane number 3 (i=3)	100	50	250	2.5
Remaining theoretical lanes (i>3)	0	0	0	2.5
Remaining area	0	0	0	2.5

Table 10 Correction factor number of lorry passages.

Number of lorries per traffic lane per year	α_q
2.000.000	1.00
200.000	0.90*
20.000	0.80*
2.000	0.70*
200	0.60*

*The correction factors smaller than 1 may only be applied if this has been agreed upon with the appropriate authority.

Load model 2

This load model consists of a single axle load that must be applied at a random position in the carriageway. The size of this axle load amounts to:

$$Q_d = \gamma_Q \cdot \beta_q \cdot Q_k$$

In which:

Q_d Calculation value changeable load

γ_Q Partial safety factor changeable load

β_q Correction factor

Q_k Characteristic value changeable load

For Q_k 400 kN must be used. This already includes a dynamic magnification factor. The correction factor β_q is considered equal to α_q.

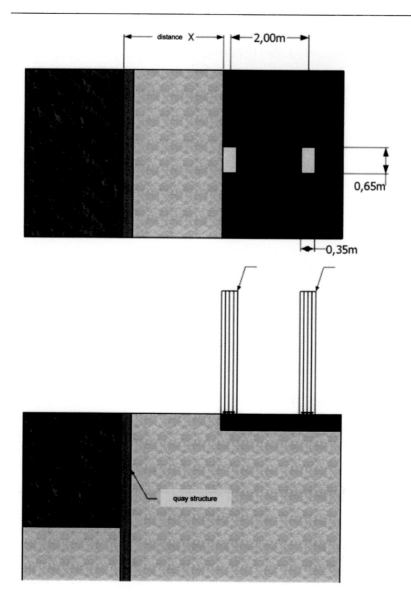

Figure 4-2 Schematic presentation of axle system in conformity with load model 2.

Table 11: Loads prescribed axle system load model 2.

Position	Concentrated axle load tandem system	Concentrated wheel load tandem system	Load resulting from wheel pressure per surface b*l (b=0,65; l=0,35)	Equally divided load per surface
	$Q_{l;k}$ [kN] *	$F_{l;wheel;k}$ [kN] *	$q_{l;wheel;k}$ [kN/m²] *	$q_{l;k}$ [kN/m²] *
Random position on theoretical lane	400	200	880	0

* This concerns characteristic values.

Direct traffic load

The carriageway behind an abutment or quay structure must be calculated in conformity with the load models described above. The axle loads of a leading tandem system must be spread in conformity with the three-dimensional tension spread according to Boussinesq. The combination of the Boussinesq method for the neutral pressure and Culmann for the active and passive pressure is often used and has been built-in in various mathematical programmes.

Most mathematical programmes calculate with a surcharge in kN/m² per running metre perpendicular to the cross-section in a 2D situation. The occurring spread is only calculated in the 2D field across the height of the wall, however, not across the scope of the wall. In regard to figure 4-3, the divided load must be manually determined by means of horizontal spread of the load in the top view up to the wall. A spread of 45° can be employed for the pavement.

This allows an equivalent evenly divided surcharge (q in kN/m²) to be calculated from, for instance, 4 wheel loads (F) across a width of the original load (D) that is representative for the calculation of a 2D situation:

$$q = \frac{4 \cdot F}{D\,(S + 2D')}$$

The influence of a surcharge or random ground level can of course also be calculated with the finite element method. These calculations are especially performed when it is necessary to know the accurate deformations of the structure or the surrounding area.

It is recommended to apply the previously described load models, but in situations in which deformations are less important, a simpler approach suffices. In this approach, load models 1 and 2 can be replaced by an equal evenly divided load that is equally divided across a rectangular surface with a size dependent upon the spread in the surface area. Please refer to chapter 6, Eurocode 7 [32], for the spread in the surface area.

95

Horizontal traffic load

When testing a structure in terms of horizontal traffic load, a distinction must be made between situations in which transition slabs are used and situations in which transition slabs are not used. Especially in case of abutments and sharp bends in a road near a quay structure, a breaking or acceleration force must be taken into account resulting from vehicles breaking or accelerating. When a stransition slab is present, the following applies:

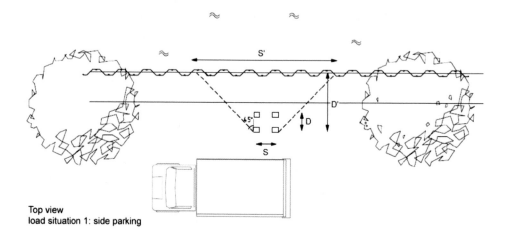

Top view
load situation 1: side parking

Figure 4-3 Load spread on a quay wall.

$$Q_d = 0{,}6 \cdot \gamma_Q \cdot \alpha_q \cdot Q_{1;k}$$

In which:

Q_d Calculation value changeable load.

γ_Q Partial safety factor changeable load.

α_q Correction factor.

$Q_{1;k}$ Characteristic value changeable load theoretic lane number 1.

This force is deemed to work simultaneously with one or multiple axle loads of load model 1 and with the ground pressure on the active side of the quay structure.

If no transition slabs are present, no horizontal load needs to be calculated unless the quay structure is part of the carriageway. In that case, the abovementioned break/acceleration force

96

must be calculated together with one axle load of theoretical lane 1. This acceleration force must be assumed as being parallel to the traffic lane.

4.2.4   Load caused by tree roots

In many locations in Dutch city centres, trees are present directly alongside quay structures. These hard structures form a growth limitation for the roots due to which the tree cannot anchor properly, increasing the chance of it being blown down. The roots in the ground can generate significant loads on the quay structure during a storm. Additionally, the root growth can also cause extra loads over time. This section provides a recommendation on how to deal with such loads. The weight of a tree must be considered as a permanent load. In urban areas, the direct wind load against a quay structure is negligible. However, depending on the running length, the wind can have an important influence on the root load against the quay wall. Wind load is variable and must be considered as a changeable load. The location of a fallen tree in relation to the anchorage of the quay structure must be calculated as a special load, if applicable.

The presence of trees within the area of influence of a quay structure consisting of a single anchored sheet pile wall increases the chance of one or multiple failure mechanisms. We distinguish an increase of the load on the quay and possible anchorage as a result of:
- the weight of the tree;
- expansion load due to the root system on the quay or anchorage structure;
- wind load on the tree that is transferred to the surface via the root system;
- the tree being blown down creating a scouring hole.

A number of aspects are also of importance with regard to management and maintenance. The accessibility of a quay structure for inspection and maintenance can be relatively complex due to the presence of trees, and the costs of conducting maintenance activities are higher. Moreover, additional provisions must often be included in the quay structure in order to preserve trees. In the case of renovation or new development, it is advised to place trees at some distance from the quay structure, at least 2.5 metres. This distance is smaller for smaller trees.

Quantification tree load

In order to determine the weight of the tree and the wind load on the tree, the height of the tree and the size of the crown must be known. Table 12 includes a number of indicative quantities for a number of tree types.

Table 12 Indicative quantities per type of tree.

		Oak	Linden	Poplar
Tree height	h_b	15 m	15 m	20 m
Crown diameter	d_k	10 m	10 m	10 m
Trunk diameter	d_s	0.4 m	0.4 m	0.4 m
Trunk height	h_s	5 m	5 m	5 m
Crown height	h_k	10 m	10 m	15 m
Weight	G	20 kN	15 kN	15 kN

The transfer of the weight and wind load on the surface takes place via the root system. For every type of vegetation, the root system consists of a so-called bearing and nourishing part. The first consists of thicker roots; the vegetation derives its stability from this part. The second part consists of finer roots (root hairs) necessary for the uptake of nutrients needed for growth and blossoming. It is difficult to clearly indicate the extent of both parts of the root system. The development of the root system is dependent upon a number of factors, including tree type, location and depth of the groundwater. Roots of the most common tree types in the Netherlands develop less well below the groundwater level. In practice, it has been shown that when trees are blown down, the roots generally snap at 1 to 2 metres from the trunk.

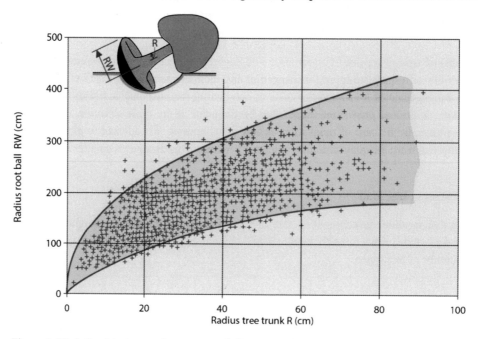

Figure 4-4 Relationship tree trunk versus root ball.

98

Mattheck and Wessolly have conducted studies into the dimensions of a scouring hole in the ground caused by a fallen tree. It is immediately noticeable that Mattheck takes significantly larger dimensions of a scouring hole in the ground than Wessolly does. Both relationships have been presented in figure 4-5. It is difficult to indicate which is true in this case. For older, but especially for younger trees, Mattheck is too high. Wessolly is slightly too low, because a correction must be added for the trunk diameter on ground level compared to that at a height of 1.3 metres. In consultation with an expert, a correction factor of 1.2 to 1.5 metres could be used, based on the type and the developmental stage of the tree. For trees that have fallen because roots were also decayed, the line of Wessolly provides an approximation which is closer to the actual situation. Although Mattheck's method gives an overestimation for the relatively young and old trees, this relationship is used as the upper limit.

The figure below can be used to determine the diameter of the root system. The radius rw of this circular cross-section is considered to be equal to half the diameter of the root ball. This value must be verified by an expert.

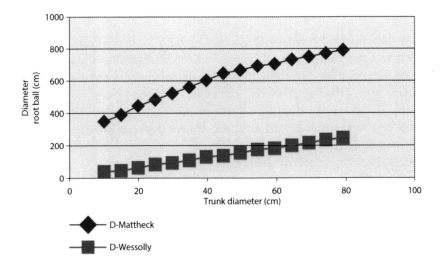

Figure 4-5 Trunk diameter versus root ball diameter.

It is assumed that the weight of the trees is born equally by the bearing part of the root system, resulting in a circular surcharge q:

$$q_{eg;rep} = \frac{G_{rep}}{\pi\, r_d^2}$$

and

$$b_{eg} = \sqrt{\pi}\, r_d$$

For schematising the sheet pile calculation, this circular load may be replaced by an equally large, evenly divided load across a rectangular surface with sides equal to √πrd. The spread of the load in the surface can be determined in conformity with the three-dimensional stress spread of Boussinesq.

Wind load

The extent of the wind load is mainly determined by the size, transparency and height of the crown and is a changeable load. There are no fixed mathematical rules for the calculation of wind load on trees. Currently, such rules are being developed for wind load on trees on primary water-retaining structures. Therefore, Eurocode 1, parts 1-4 [19], are used at the moment. This provides values for calculating the wind load for building structures, depending on the height above ground level and the geographical location. An important point of attention in determining the wind load, is the influence of local buildings. These buildings can significantly reduce the wind load.

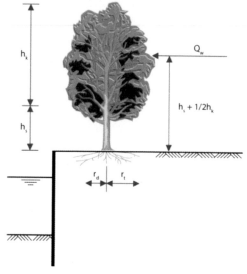

Figure 4-6 Wind load on tree.

Assuming that the surface affected by wind load is equal to the crown surface, the resulting horizontal wind load, with leverage point half way up the crown (trunk height + 0.5 times the crown height), can be determined as follows [38]:

$$Q_{w;rep} = c_s \cdot c_d \cdot c_f \cdot q_p(z) \cdot A_k$$

In which:

$Q_{w;rep}$	Resulting wind load
A_k	Frontal surface crown
c_s, c_d	Structure factor (for instance 0.3)
c_f	Force coefficient
$q_p(z)$	Thrust at height z

This wind load is transferred to the surface via a horizontal shear force to the amount of Q_w and a vertical torque M_w. For the vertical torque M_w the following applies :

$$M_{w;rep} = Q_{w;rep} (h_s + 0{,}5h_k)$$

In which:

$M_{w;rep}$	torque
$Q_{w;rep}$	Resulting wind load
h_k	Crown height
h_s	Trunk height

Depending on the flexural rigidity of the quay structure and the spring stiffness of the anchorage, the shear force $Q_{w;rep}$ can be spread across a larger width. With an analysis of the redistribution capacity of the quay structure concerned, insight may be acquired into the measure of spread. To simplify this, the force can also be translated directly into an increase of the anchor force to be delivered by the bordering anchors. It is assumed here that a tree is positioned midway between two anchors and burdens each bordering anchor with half the horizontal shear force ($0.5\ Q_{w;rep}$). The vertical torque $M_{w;rep}$ on ground level consists of a pressure strength ($Q_{v;rep}$) and an equally large tensile strength ($Q_{v;rep}$) at a mutual distance of $x_{pressure} + x_{tensile}$.

$$M_{w;rep} = Q_{v;rep} \left(x_{tensile} + x_{pressure} \right)$$

For the pressure strength, the centre of gravity of half the bearing part of the root system, with a radius of r_d, is taken as leverage point $x_{pressure}$. The same reasoning cannot simply be applied to the leverage point of the tensile strength, hereafter referred to as $x_{tensile}$. After all, the ground cannot absorb tension. This means that the tensile strength must be delivered by the ground weight hanging on to this part of the root system. The size of this ground segment is determined by the effective thickness of the segment and the unit weight of the ground. The part that must deliver tensile strength can be pictured as half a circle with radius r_t:

$$x_{pressure} = \frac{4r_d}{3\pi}$$

$$x_{tensile} = \frac{4r_t}{3\pi}$$

$$Q_{v;rep} = 0{,}5\,\pi r_t^{\,2}\,\gamma_s\,h_m$$

In which:

$Q_{v;rep}$	Resulting pressure/tensile load (kN)
h_m	Effective width (assumption 1.0 m)
γ_s	Unit weight ground (kN/m³)

$Q_{v;rep}$ and r_t can be solved by means of the above relationships. Subsequently, the strip load can be determined as follows:

$$q_{pressure;rep} = \frac{Q_{v;rep}}{0{,}5\pi r_d^2}$$

$$q_{tensile;rep} = \frac{Q_{v;rep}}{0{,}5\pi r_d^2}$$

In which:

$$b_{pressure} = 0{,}5\sqrt{\pi}\,r_d$$

$$b_{tensile} = 0{,}5\sqrt{\pi}\,r_t$$

The modelling described here is in many cases conservative, due to which the requirements arising from valid norms are not met. Optimisation can often lead to a stronger load reduction. Think of an accurate determination of the coefficient resistance. Furthermore, an accurate determination of the root ball and the present taproots also greatly affect the load of the quay wall. A tailored solution can also often provide more accurate and realistic results, where an integral modelling by means of a programme based on the finite element method is performed.

A frequent situation in urban areas is the presence of trees that were planted just a short distance from the quay structures. These trees create a dual load on quay walls. On the one hand, the growing root system causes a certain root pressure against the wall. On the other hand, wind load on the crown results in a reaction force being delivered in the ground mass directly situated behind the quay structure and possibly against the quay structure itself. For a more detailed description of these loads, please refer to section 4.2.4. The suggested load scheme is a first, conservative estimation of the load effects.

When a sensitivity analysis shows that the loads mentioned could have a significant influence on the stability of the quay wall, it is recommended to use an integral finite element model. Such a model provides a better representation of the actual load transfer to the surface and the indirect influence on the quay structure. A finite element programme that is often used in the Netherlands is PLAXIS. Additionally, effects of possible measures can be better investigated/simulated with this programme. For the urban quay wall alongside the Prinsessegracht in The Hague, a comparable inspection was performed for a representative cross-section, see figure 4-7 a and b. This concerns a situation of a concrete quay wall with brickwork at the front, founded on wooden piles. Also, a wooden seepage screen has been applied. By means of finite element calculations, the current safety level of the total structure has been mapped, including tree load and traffic load. The size of the root system is taken from ground level up to the highest groundwater level and the root system is schematised as a reinforced ground segment by assigning cohesion to this ground segment. The tree load is modelled as a set consisting of a horizontal line load, a vertical line load and a moment of leverage to one metre below ground level. As the topside of the quay structure concerned was subject to horizontal deformations, the effect of applying a tilted anchorage was investigated to once again restore the structure to the desired safety level.

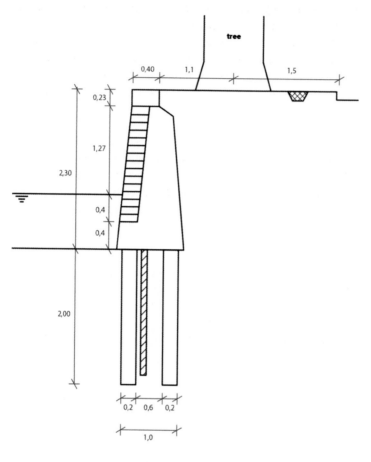

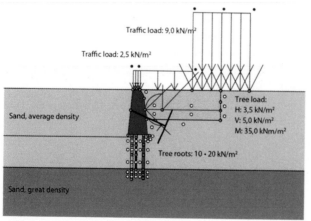

Figure 4-7a & b Soil structure and loads.

It must be noted that deformation capacity/flexural rigidity of the tree have not been taken into account. Consequently, the actual stress on the surface will be less. The constructor must also take into consideration the flexural rigidity of the urban quay structure in testing and design. As the flexural rigidity of the quay structure increases, the loads are spread across a greater length, which in turn will lead to a decrease in line load per running metre of quay. Nevertheless, tree load must be taken into account in the design and testing of quay structures. As the distance of the tree to the quay increases, the influence of the tree on a quay structure will reduce significantly due to tension spread in the ground.

Practically all municipalities have indicated that trees in the immediate vicinity of quay structures have caused damages. The wind load in urban areas can be significantly reduced by the presence of local buildings. On the other hand, the wind speed may be significantly higher, due to local tunnelling.

Fallen tree

The least favourable situation for the stability of the anchorage occurs when the tree is blown down and a hole arises in the ground in front of an anchor screen or deadman anchorage, for instance. This can be considered as a special load. Observations of fallen trees (oak and linden) show a maximum hole depth of 1.0 to 1.5 metres arising across a surface equal to the bearing part of the root system. This hole depth and surface must be established by a tree expert. In most cases, the groundwater level can be taken as a reference. As an initial approach, a surface area of 3.0 by 3.0 metres seems to be a safe assumption (linden and oak). Depending on the location of the tree with regard to the deadman anchorage, this "scouring" must be taken into account when determining the passive resistance of the deadman anchorage.

4.3   Test method

4.3.1 *Introduction*

In this section, the test method is described, which matches the step-by-step approach of chapter 4. The goal of testing is to record the condition of the structure and especially the residual lifetime of the objects. The scope of the test is dependent upon the available data and the inspection results. For this purpose, a division can be made into three classes: a simple test, a detailed test and an advanced test.

This approach is in accordance with the Regulation Testing on Safety primary water retaining walls (VTV2006, [16]). A simple test entails that the entire structure can be approved or disapproved on the basis of simple rules. For a quay wall, the judgement "structurally safe" can be given if it can be demonstrated that the quay wall has been designed in accordance with the valid norms and guidelines, the structural components are not degraded, and the load has not increased. In many cases, the structural safety of a quay cannot be established on the basis of simple criteria and a detailed test must be conducted.

During a detailed test, it is demonstrated that every critical component of the structure possesses sufficient safety and that balance is ensured by means of common calculation models. Here, the possible failure mechanisms per type of quay wall must be inspected, see below in this section. Schematising a sheet pile wall as an elastically supported beam can provide insight into the residual capacity of the steel cross-section. Calculations must be performed per component in conformity with Eurocode 7 [32], in accordance with NEN 8700 [31] (Dutch national standards), where the partial safety factors may be reduced in conformity with the tables in appendix 3, 2 and 5, depending on the required residual lifetime and the consequences in the case of failure. When it can be demonstrated for one of the critical components that the structural safety cannot be assured, then the judgement "structurally unsafe" may be given.

For the benefit of detailed and advanced testing ,the current strength of the structural elements must be used in calculations. This strength can be determined with the aid of chapter 3.

An advanced testing takes place when no judgement can be given concerning structural safety after applying the simple and detailed testing. For testing on this level, the geometrics could be integrally inspected by means of the finite element method.

The test criteria must be recorded in the Terms of Reference. For a detailed flowchart including the various testing steps, please refer to appendix 1.

The quay structures are distinguished in four main forms in case of testing:
- type 1: gravity wall on spread foundations.
- type 2: gravity wall on pile foundations.
- type 3: L-wall on pile foundations.
- type 4: sheet pile wall from steel or concrete.

Failure mechanisms and type of retaining wall

As part of setting up a test method, the failure mechanisms of the quay structures have been defined and discussed in more detail.

The most important geotechnical failure mechanisms have been established to test the current situation of a quay wall. These mechanisms have been established in line with the Eurocode. When a failure mechanism or deformation in the soil structure or the subsurface occurs and this is of such a magnitude that it causes a failure mechanism in the soil structure, this is referred to as an ultimate limit state. In conformity with Eurocode [17], this ultimate limit state has been defined as the state of the structure when it is only just capable of fulfilling its function.

Eurocode 7 [32] distinguishes between the following ultimate limit states:

- GEO: 'failure or large deformation of the subsurface, where the strength of the soil or rock formation contributes significantly to the resistance'(bearing capacity foundations and retaining structures, failure of the soil, soil loaded structures).
- STR: 'internal failure or large deformation of the structure or parts of it, including spread foundations, pile foundations or wall foundations, where the strength of the construction materials must contribute significantly to the resistance' (failure of the structure).
- EQU: loss of balance of the structure or the subsurface, taken as a rigid whole, in which the strength of the construction materials and the subsurface do not significantly contribute to the resistance.
- UPL: 'loss of balance of the structure or subsurface resulting from lifting due to water pressure or other vertical loads'(lifting structure, failure of tension elements), in which the resistance of the subsurface may be a contributing factor.
- HYD: 'hydraulic heave, internal erosion and erosion caused by concentrated groundwater flow in the subsurface resulting from hydraulic gradients' (piping, bursting of the bottom).

On the basis of these general limit states, the (geotechnical) failure mechanisms can be established per type of quay wall. Additionally, failure of functionality due to too large deformations (in the usability limit state) may occur, also see the next section.

A well-known failure mechanism that is not geotechnical, but is highly relevant to old quay structures, concerns the bulk density of the structure. Due to sand/earth washing away, hollow spaces are created that could lead to failure of the structure. This failure mechanism cannot be calculated, but it is often the result of degradation of the structure.

Failure mechanism retaining wall on spread foundations (type 1)

A gravity wall on spread foundations may consist of a solid structure of stacked brickwork or basalt columns, a structure consisting of caissons, or a concrete structure, possibly with a brickwork or basalt covering. The structure is directly founded on the subsurface, usually on a (reinforced) concrete floor. Additionally, the L-walls on spread foundations are also categorised as this type of quay wall. In conformity with Eurocode 7, these are practically equal to gravity walls on spread foundations, where the ground segment above the foot of the L-wall can be considered 'dead' weight (with matching safety approach).

The most important failure mechanisms are as follows:

1. Exceeding the vertical bearing capacity of the subsurface (GEO):
   a. exceeding the vertical bearing capacity in the drained situation;
   b. exceeding the vertical bearing capacity in the undrained situation;
   c. exceeding the vertical bearing capacity due to punch (drained or undrained);
   d. exceeding the vertical bearing capacity due to undrained squeezing sideways.

2. Exceeding the horizontal bearing capacity of the subsurface (GEO):
    a. horizontal shearing in the drained situation;
    b. horizontal shearing in the undrained situation.

3. Exceeding the general stability (GEO).

4. Exceeding the tilt stability (forces with large eccentricity) (EQU).

5. Failure of the structure (exceeding the structural strength) (STR).

6. Failure of the structure due to large displacement of the foundation (STR).

Failure caused by internal erosion or piping (HYD). The failure mechanisms for type 1 have been schematically presented in figure 4-8.

In addition to the testing of the abovementioned failure mechanisms in the ultimate limit state, the functional requirements in the usability limit state must also be tested. This limit state also leads to a (functional) failure mechanism when:

7. Displacements are of such a magnitude that the allowable values are exceed when:
    a. functionality is compromised;
    b. esthetical requirements are exceeded;
    c. these lead to unallowable deformation of structures in the immediate vicinity.

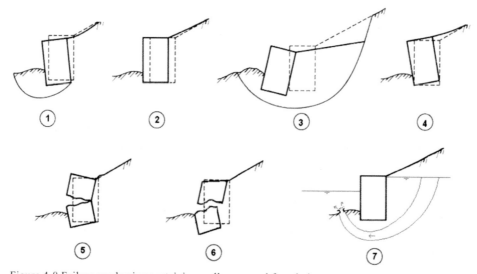

Figure 4-8 Failure mechanisms retaining wall on spread foundations

Failure mechanisms retaining wall on pile foundations (types 2 and 3)

A retaining wall on pile foundations (this is type 2) can consist of a solid gravity wall of stacked brickwork or basalt columns, or a concrete structure possibly with brickwork or basalt covering, placed on a (reinforced) concrete floor which, in turn, is founded on piles. The piles could be wooden, concrete or steel tubular piles where at least one row has been placed as much as possible at an inclination to transfer the horizontal loads (in connection with the vertical load) to the subsurface as axial pile force. The failure mechanisms also apply to relieving platforms on pile foundations and L-walls on pile foundations (quay wall type 3). The most important failure mechanisms are as follows:

1. Exceeding of the vertical pile (pressure) bearing capacity (GEO).

2. Exceeding the tension bearing capacity (GEO/UPL):
   a. exceeding for several piles;
   b. exceeding for pile groups, including exceeding pull-out cone weight.

3. Failure of the surface due to a horizontal load on the pile foundation (GEO).

4. Exceeding the general stability (GEO).

5. Structural failure of the retaining wall (STR).

6. Structural failure of the piles due to pressure, tension, bending, buckling or shearing (STR).

7. Collapsing of the structure due to large displacement of the foundation (STR).

8. Collapsing caused by internal erosion or piping (HYD).

The failure mechanisms for types 2 and 3 have been schematically presented in 4-9.

In addition to the testing of the abovementioned failure mechanisms in the ultimate limit state, the functional requirements in the usability limit state must also be tested, as mentioned for type 1.

Failure mechanisms sheet pile wall (type 4)

Sheet pile walls actually concern retaining structures fixed in the ground. Steel or concrete sheet pile walls can be applied, but also wood and synthetic materials for lighter structures such as small retaining structures. The sheet pile walls can be anchored or free standing, covered with brickwork and hanging structures or fitted with a concrete capping beam.

The following failure mechanisms apply to sheet pile walls:

1. Exceeding the strength of the wall due to flow/breach extreme fibres due to combination of torque, shear forces and axial forces possibly with corrosion (STR).

2. Exceeding the passive horizontal soil resistance (exceeding bearing capacity subsurface, leading to heaving) (GEO).

3. Exceeding the vertical bearing capacity of the subsurface at bearing sheet pile walls (GEO).

4. Exceeding the general (macro) stability (GEO).

5. Collapsing of the structure due to rotation or translation of the wall or parts of it (STR).

6. Collapsing of the anchorage (STR/GEO):
   a. collapsing of anchor rod due to flow or breach;
   b. exceeding the tension bearing capacity of the grout body (slip criterion);
   c. exceeding due to lateral load due to settling soil on anchor tie rod;
   d. exceeding soil resistance anchor wall.

7. Exceeding the Kranz stability (in case of anchorage) (GEO/EQU).

8. Failure due to hydraulic heaving and piping (HYD/UPL)

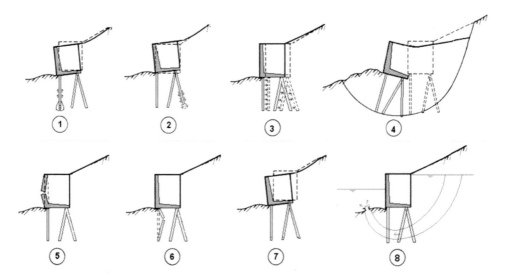

Figure 4-9 Failure mechanisms retaining wall on pile foundations.

The failure mechanisms for type 4 have been schematically presented in figure 4-10.

In addition to the testing of the abovementioned failure mechanisms in the ultimate limit state, the functional requirements in the usability limit state must also be tested. This limit state also leads to a (functional) failure mechanism when:

9. Displacements are such that these exceed the allowable values when:

110

a. functionality is compromised;
b. esthetical requirements are exceeded;
c. these lead to unallowable deformation of structures in the immediate vicinity;
d. undesirable 2nd order effects may occur;
e. unacceptable leaking and/or transport of particles through or underneath the wall may occur.

It must be noted that failure limit states may occur in drained and undrained situations in case of poorly permeable soil and that these must then also be analysed.

Non-geotechnical failure mechanisms

Non-geotechnical failure mechanisms especially relate to the state and the functioning of the structure. For example:
- degeneration of the strength of the structure;
- degeneration due to growth on the structure;
- soil density of the structure;
- failure due to tree roots;
- biological and environmental effects, influence of plant growth;
- failure due to exceptional loads such as impact load during parking.

Summary of failure mechanisms

The failure mechanisms have been summarised per type of quay wall in table 13.

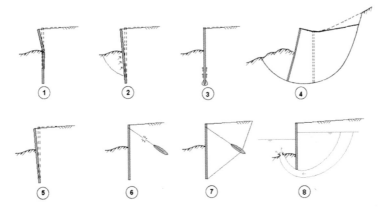

Figure 4-10 Failure mechanisms sheet pile wall as quay.

111

Table 13 Failure mechanisms per type of quay wall.

Failure mechanism (geotechnical and structural)	Limit state	Retaining wall type 1	Retaining wall type 2	Retaining wall type 3	Retaining wall type 4
Vertical bearing capacity subsoil	GEO	x			
Vertical pile bearing capacity (pressure)	GEO		x	x	x
Vertical pile bearing capacity (tension)	GEO/UPL		x	x	
Horizontal bearing capacity subsoil	GEO	x	x	x	x
Tension resistance anchorage[1]	GEO/UPL				x
Macro stability	GEO	x	x	x	x
Tilt stability	EQU	x			
Kranz stability	GEO/EQU				x
Structural strength wall / retaining wall	STR	x	x	x	x
Structural strength piles	STR		x	x	
Structural strength anchorage[1]	STR				x
Structural strength remaining elements[1]	STR				x
Failure due to large displacement	STR	x	x	x	x
Internal erosion	HYD	x	x	x	x
Seepage & piping	HYD	x	x	x	x

1) could also apply to type 1, 2 or 3 if anchorage, soil nailing or soil reinforcement has been applied.

4.3.2 Reference period and residual lifetime

The term 'residual lifetime' must be properly distinguished from 'reference period'. The residual lifetime is the assumed period during which an existing or reconstructed structure or part of this can be used for the intended goal. Within this period, the minimum safety level may not be exceeded. The reference period is the time period chosen and used as principle for statistical value determination of changeable loads and possible exceptional loads. Therefore, the reference period does not have to be equal to the residual lifetime.

For existing structures, the period that the structure must still be functioning is often shorter than the standard reference period of 50 years when designing new structures. By employing a shorter reference period, an existing structure will not be disqualified as quickly. The changeable loads can be determined for a shorter period of time, for instance, so that these will turn out with a lower value. Additionally, the degradation of the structure for a shorter reference period will be less.

If a structure does not meet the requirements connected to the intended residual lifetime of (parts of) the structure, an inspection must take place as to whether the disqualification level has been exceeded. The disqualification level is the legal minimum level of structural safety, the exceeding of which will give rise to notification and enforcement by the authorities.

It is noted in NEN 8700 [31] that, in case of a disqualification assessment, irrespective of the zoning and size of the structure, a residual lifetime of 1 year must be assumed. For the reference period for the determination of changeable loads, the period equal to the assumed residual lifetime, being 1 year must be assumed for evaluating whether or not the performance level of a structure exceeds the disqualification level for consequence class CC1a, (this concerns the safety at any one time). For consequence classes CC1b, CC2 and CC3, a reference period of at least 15 years must have been used, even though the (assumed) residual lifetime for the evaluation of the strength of the structure is 1 year.

Note that, in NEN 8700, a distinction is made between class CC1a and CC1b. This distinction entails that, for CC1a, the loss of human life is excluded, whereas for CC1b, the danger of loss of human life is small.

4.3.3 Standards

Urban quay walls can be part of a primary or secondary water retaining structure. Much more stringent safety requirements apply to primary water retaining structures than is the case for quay walls that are not part of a primary water retaining structure. Please refer to the Guideline Civil Engineering Structures [37] and the reports of Deltares regarding 'Coordination Guideline Engineering Structures and Eurocode' for the assessment of such structures. For secondary water retaining structures, however, less stringent safety requirements apply to the water retaining function compared to the safety requirements for 'normal' building structures.

In such situations, the quay structure can be assessed as a building structure in conformity with the Eurocode.

In the earlier legislation NEN 6700, three reliability classes were distinguished, namely with required reliability indexes of 3.2, 3.4 and 3.6, belonging to reliability classes 1, 2 and 3 respectively. The unwritten policy for engineering structures managed by the Dutch department of Infrastructure is that primary water retaining engineering structures fall into the highest reliability class. Under this earlier legislation, this implies that a safety index of at least $\beta N = 3.6$ was required, where the standard reference period of N=50 years was used.

In the Eurocode [17], three reliability classes are distinguished with required reliability indexes of 3.3 (reliability class RC1), 3.8 (RC2) and 4.3 (RC3). The earlier reliability classes 1 and 2 currently belong to reliability class RC1, reliability class 3 (largely) under RC2, and a new consequence class RC3 has been created for building structures of which failure would lead to highly significant societal impact. Most quay structures are subject to reliability classes RC2 or RC3. A minimum reliability index of 3.8 or 4.3 respectively is required, related to a reference period of 50 years in the case of new structures.

Norm NEN 8700 [31] was established especially for the assessment of existing structures. This norm must be applied to the assessment of existing structures and reconstruction projects. The norm was established because the evaluation of an existing structure deviates in a number of essential points from that of new structures.

Firstly, the deliberation for the desired safety level in relation to cost effectiveness is generally different. After all, increasing the safety level or realising the level of a new structure is often accompanied by relatively higher costs in a reconstruction situation than is the case in the design stage. Additionally, the period during which a structure must still be functioning is often shorter than the standard reference period of 50 or even 100 years for existing engineering structures. This has already been discussed in the previous section. Finally, there is the possibility of obtaining more information on a building structure via measurements.

NEN 8700 [31] prescribes a minimum reference period of 15 years and a minimum reliability index of 1.8, 2.5 and 3.3 respectively for existing structures with reliability class RC1, RC2 and RC3. These reliability indexes are significantly lower than those applicable to new structures, namely a reliability index of 3.3, 3.8 and 4.3 for RC1, RC2 and RC3 respectively. The adjusted reliability levels, in the form of decreased beta values, have been indicated for existing structures in sections 5.2.1 and 5.2.2.

In order to assess an area of existing building structures, a value lying between the level of a new structure and the disqualification level must be used for existing building structures, with the reconstruction level being the maximum. The minimal requirements for the construction level are described in chapter 6.

A recommendation for reassessment of an area of existing building structures is to assume the reconstruction level; in case of disproportional expenses, a lowering to the disqualification level could be allowed within the area in incidental cases. Think of listed building structures, for instance. For individual building structure owners, the advice is to only allow the disqualification level if measures to meet the reconstruction level can be taken in the foreseeable future. The foreseeable future refers to a time period of a number of years, depending on how much the reconstruction level has been exceeded. For further details, please refer to [31]. The disqualification level may only be accepted for structures that are permanently loaded. Additionally, the behaviour or the structure must be monitored for as long as it fails to meet the reconstruction level.

For retaining walls on spread foundations or pile foundations, type 1 up to 3, according to the Eurocode [17], the differentiation in reliability index occurs through the differentiation of load factors between the different reliability classes. For this type of structure, the material factors for all reliability classes are equal. In NEN 8700 [31], alternative load factors are given for existing structures, compared to the values applicable to new structures. The load factors from NEN 8700 are lower than those applicable to new structures. However, these load factors have been derived for buildings and bridges. The Dutch norm NEN 8707 is being developed for geotechnical structures. Presently, there are no load factors available for the assessment of existing geotechnical structures. Awaiting NEN 8707, it is currently advised to apply the load factors derived for the assessment of existing structures, whilst being aware of situations and loads that do not occur in buildings. Think of soil pressures, for example. In these cases, one must act in accordance with the normative parts of the Eurocode, appendices B and C, including the national appendices.

For sheet pile walls, quay walls type 4, the distinction in reliability index is made by differentiating both material and load factors, according to the Eurocode [17]. It must be noted that this distinction especially relates to the soil parameters. NEN 8700 offers no guidelines for a distinction in material factors, as this norm was primarily developed for buildings and bridges. For geotechnical structures, norm NEN 8707 is being developed. Presently, there are no load factors available for the assessment of existing geotechnical structures.

Awaiting NEN 8707, it is advised, for this type of structure, to assume the material factors from CUR 166 and the Eurocode derived for a certain reliability index for a reference period of 50 years. This is a conservative length of time compared to a reference period of 15 years. For existing structures with reliability class RC1, RC2 and RC3 respectively, the material factors can be utilised that are also applied to new structures in CUR class I, RC1 and RC2 respectively. In incidental cases, the material factors that are also applied to new structures in a SLS calculation, CUR class I and RC1 respectively can be applied. As previously noted, incidental cases refer to listed building structures or other special structures which are permanently loaded and monitored.

Table 14 of chapter 6 indicates how the different reliability classes relate to each other.

4.3.4 Length effects

Length effects must be taken into account in the case of structures of which failure of a cross-section leads to failure of the structure as a whole. The length effect does not play a role for quay structures that do not have a primary or secondary water retaining function because failure only results in local damage in such cases. The length effect must be taken into consideration if the quay structure is part of a primary or secondary water retaining structure. In order to calculate the length effect for primary and secondary water retaining structures, please refer to the Guideline Engineering Structures [37].

4.3.5 Proven strength

In general

Many structures have been functioning properly for dozens of years. Usually, degradation of structural components will have occurred (of which the inspection can be very expensive). Moreover, the load on these quays has often changed over the years. The traffic load in old city centres has often increased, for example. However, in old city ports, the quay load has usually decreased: storage of goods used to take place here, whereas nowadays, the quay is used as a boulevard or for parking cars. In the latter case, it is possible to apply a method of 'proven strength'. It is, however, necessary to have insight into the historical loads (surface, static/dynamic, duration, etc.), interim alterations to the structure and the current state and behaviour of the structure.

If, during an assessment, it is demonstrated that the quay wall does not meet new guidelines, it can be approved on the basis of proven strength, if:
- it has been able to withstand decisive loads (especially retaining height, extreme (ground) water levels and surcharge) in the past;
- no significant damage has occurred in this extreme situation;
- the quay structure has not undergone (disadvantageous) alterations. Think of degradation resulting from corrosion;
- all this information can be established with sufficient reliability.

Correction factor approach

In practice, it will rarely occur that all these conditions are met. The proven strength method can still be applied, but additional mathematical analyses are required. This approach is called the 'correction factor approach' in the Technical Report Actual Strength of Dikes [33]. This section provides a brief summary of this method. For more detailed information, please refer to the abovementioned reference.

The proven strength approach for quay structures is dependent upon which failure mechanism is not met. For most tests, the following must apply:

$$\text{F.S.}_{d;0} = F_r / F_s \geq 1.0$$

in which:

$\text{F.S.}_{d;0}$ Stability factor for the (current) leading situation

F_r Calculation value of strength

F_s Calculation value of load

If this safety factor does not suffice, but the structure has withstood a large load without significant damage in the past, there is reason to assume that the actual strength is greater than the calculated safe estimation of this strength.

If the structure has not undergone any disadvantageous alterations since this historical situation, a correction factor can be determined according to:

$$\gamma_{cor} = 1 / \text{F.S.}_{his}$$

In which:

γ_{cor} Correction factor proven strength

F.S._{his} Stability factor for the historical situation

Please note that this method will only lead to a correction factor larger than 1.0 if the safety factor for the historical situation is smaller than 1.0. The corrected safety factor can subsequently be calculated according to:

$$\text{F.S.}_{d;0;cor} = y_{cor} \times \text{F.S.}_{d;0}$$

in which:

$\text{F.S.}_{d;0;cor}$ = Corrected stability factor for the (current) leading situation

y_{cor} = Correction factor proven strength

$\text{F.S.}_{d;0}$ = Stability factor for the historical situation

This can subsequently be tested against the norm. This approach must be followed for every failure mechanism that has not been assessed as sufficient. In order to apply the 'correction factor approach', a minimum of the following information must available:

- geometrics (ground level, retaining height, water levels) at the location of the influence area of the structure;
- external loads on the structure, such as water levels and surcharges;
- information on the behaviour of the structure during the extreme situation;
- structure of layers at the location of the structure;
- soil characteristics per layer (real average values versus low characteristic values);
- water tensions at the location of the structure.

The abovementioned information often includes uncertainties. It is important to distinguish between systematic and non-systematic uncertainties.

For a systematic uncertainty, it can be assumed that the uncertainty is constant. This might apply for the layered structure, for instance. This will generally not have changed much, but is uncertain to a greater or lesser extent. Such uncertainty must be covered in the analyses by considering various scenarios. A practical method is to determine the correction factor with a conservative, average and optimistic scenario. It is possible that a less conservative estimate of the layered structure will be of more influence in the lowering of the correction factor than in increasing the safety factor for the current decisive situation.

Additionally, non-systematic uncertainties may occur. These are uncertainties in influence parameters that can differ greatly per situation. This is the case of the surcharge or the influence of precipitation. These uncertainties must be calculated by assuming favourable rather than conservative schematising in the calculation of the historical situation. In the test calculation, a load must be included, unless this can be excluded. Conversely, for the historical circumstances, this must not be included, unless it can be guaranteed. The reverse is true for the strength. By taking into account uncertainties in this way, a sufficiently safe estimation for the correction factor can be obtained. When, in theory, 0% degradation of the structure occurs and there is 0% chance of a higher load than the structure has historically survived, it has been proven that the strength suffices in a relatively simple way. Unfortunately, this is more complex in practice.

It is especially difficult to estimate the degradation of the structure and the structural parts. This does not only concern the current state, but also the state at the time of the historical load and the state for the reference period to be tested. This is a non-systematic uncertainty. Because a predication of degradation (from the current state until the end of the reference period) is surrounded by uncertainties, the as-built situation (without degradation) is assumed for the state at the time of the historical load. The influence of possible interim renovations/ reconstructions must be included in the analysis.

Please note that this method has only previously been applied to macro stability of water retaining structures. The method needs to be further developed for quay structures. Presently, the method is only suitable for failure mechanisms that can be expressed in a safety factor on the basis of a deterministic approach. This is a level 0 method, see section 5.2.3. A semi-probabilistic approach is also possible in theory, but insufficient information is generally available for such an analysis. Therefore, no attention is paid to this option in this publication.

Another aspect that must be considered is the length effect. For water retaining structures, proven strength is applied to historical successfully retained water levels (without damage to the water retaining structure). The historical load for quay structures usually consists of a high surcharge, that was often only present on certain parts (the entire quay length was not loaded). A substantiation of the strength is in many cases only possible if it can be substantiated that, at the time of the historical load, the entire quay length:
- was in the same state;
- the quay has not been altered in the meantime;
- the quay displays the same behaviour across the entire length;
- the soil composition is similar across the entire length;
- the quay structure has the same design across the entire quay length.

To calculate the proven strength, it is also recommended to obtain insight into dredging activities or erosion in front of the quay. Future scenarios that are based on the usage of the quay, must be taken into account. For example, certain ships can lead to more scouring (in case of missing bottom protection). Insight into past restorations is also important, especially when these were performed on a certain part.

4.3.6   Geometrics

For testing, the inspection data must be translated into key ratios that can be used in the mathematical test. Every test includes uncertainties. These can largely be decreased by performing field measurements and inspections, but residual uncertainties will always remain. For wooden pile foundations, the following uncertainties can be distinguished.
- pile tip level;
- pile configuration (distances);
- inclinations;
- diameter piles.

Inspections aim to acquire a representative view of the structure and parts, but the reality may deviate from this. Therefore, the inspection results must be statistically processed. Additionally, the extensiveness of the inspection must be coordinated with the variations to be expected, in order to create a representative view.

The water levels and rising heights to be used can also contain uncertainties. A significant portion of the load on the quay structure is formed by a possibly present hydraulic gradient across the quay structures. Leaking often occurs at old quay structures. In practice, the hydraulic gradient appears to be highly limited. Especially in renovations, it must be critically assessed whether the adjustments do not lead to stagnation in the groundwater level. It is important to monitor the groundwater level in relation to the water level in the port for the test calculations.

In doing so, both periods of extreme rainfall and dry periods must be considered.

4.3.7   Non-structural elements

In general

Non-structural elements on or in the area of influence of the quay do not have a structural function for the quay, but can (negatively) influence it. Think of the following types of objects:
- trees/plants;
- street and quay details;
- cables and pipes;
- storage facilities;
- jetties/fenderings;
- mooring facilities.

Influence zone

Objects can influence the quay structure in multiple ways:
- Load
  Objects in the active or passive zone burden or support the quay structure. This is only the case if the quay structure lies on a foundation (spread foundation). In case of deep foundations, the influence is usually non-existent.
- Foundation
  If objects have the same foundation level (both spread foundations and pile foundations) as the quay structure, the influence of foundations must be taken into account. This aspect especially deserves attention in the application of non-soil displacement pile systems.
- The influence zone of trees appears to be larger in practice than what has been theoretically established in the guidelines. The root ball is strongly related to the type of tree and the condition of the tree. This can lead to loads on the quay (static loads and effects of wind loads), damage resulting from root growth and also situations of windthrow, causing scouring.

Cables and pipes

The two most important points of attention for cables and pipes are:
- In the case of cable/pipe lead through, it is important that the lead through is soil-proof. There is otherwise a risk that the quay structure will be undermined and that hollow spaces will arise. This can be checked by means of underwater inspections.

120

- With regard to pipes, one must be aware of the fact that additional risks exist for explosion-sensitive and pressure pipes that can lead to serious damage in times of crisis. Objects can also influence each other. The uprooting of a tree can, for example, lead to damage to the underground infrastructure.

4.4   Testing

Testing existing quay structures can be conducted on three levels; simple, detailed and advanced. By means of the test results, follow-up steps can be taken. The course of the testing is presented in the flowchart in appendix 1.

4.4.1   Simple testing

For structures of which the design and management information is in order, a simple testing can be conducted. For a structure to be assessed as sufficient according to the simple testing, all conditions below must be met:
- Design calculations are available and meet the applicable legislation of that period.
- It can be demonstrated that the delivery of the structure is in conformity with the design.
- The behaviour of the structure does not deviate significantly from what could be expected according to the design (for instance, much more deformation).
- The structure is not overloaded during its lifetime.
- The boundary conditions (loads, etc.) from the design are valid and will remain valid for the residual lifetime and reference period.
- The already occurring degradation + the expected degradation during the residual lifetime and reference period of the structure are within the design values.

If all abovementioned conditions are met, the structure can be approved. In other cases, the structure must be disqualified, or further detailed testing must be conducted.

4.4.2   Detailed testing

If the criteria for simple testing cannot be met, further detailed testing can be conducted. For the detailed testing, calculations must be performed. For this purpose, the failure mechanisms as described in the previous section must all be assessed in accordance with the valid norms. In many cases, this will be the NEN 8700 series [31]. For primary and secondary water retaining structures, please refer to the Guideline Engineering Structures [37].

For detailed testing, the current strength of the materials must be calculated. For wooden foundation piles, for instance, the degradation of the wood strength must be calculated. Appendix 2 includes a guideline for these calculations.

If the requirements of the detailed testing are met, the structure can be approved. If that is not the case, the structure must be disqualified, or an advanced testing can be conducted.

121

4.4.3 Advanced testing

For the advanced testing, a proven strength analysis can be conducted. A guideline for this has been given in the previous section. Additionally, appendix E of NEN 8700 [31] contains several more guidelines to further attune the assessment of existing buildings. The following guidelines are given here, for example:

- Tightening safety factors. For instance, by determining influence factors more accurately by means of probabilistic calculations. Furthermore, in some cases the variation coefficient of the material properties or loads can be established more accurately by means of measurements.
- Adjusting the chance of failure of a structure on the basis of inspections.
- Performing test loads.

4.5 Adjusting test criteria

If the quay structure is disqualified because of the established test criteria, the possibility exists to adjust the test criteria. Think of adjusting the functional requirements, for instance. Consequently, a lower reliability class could be applied or the loads on the quay structure could be reduced. The loads could also be reduced by shortening the intended residual lifetime or by reducing the calculated degradation. After adjusting the test criteria, the test must be performed again.

4.6 Disqualification assessment

If none of the prior steps has led to an approval, the quay must be disqualified. According to NEN 8700 [31], a residual lifetime of 1 year and a reference period of 1 year for building structures in consequence class CC1a and a reference period of 15 years for building structures in consequence classes CC1b, CC2 and CC3 must be used in case of a legal disqualification level. For this situation, a calculation must be made to determine the risk profile. The risk profile is generally highly dependent upon the structural part or failure mechanism that does not meet the disqualification assessment.

4.7 Consideration

The risk profile that has been recorded in the disqualification assessment can be used to check whether repair is economically feasible, in which case the existing structure is repaired or improved, or whether a new structure must be built, in which case the existing structure is replaced. This is discussed in further detail in the next chapter.

5 Repair or new structure

5.1   Introduction

After the choice has been made to repair (reconstruct) or completely replace (new build) the existing quay structure, the structural parts concerned must be redesigned and redimensioned on the basis of the applicable reliability criteria.

In this chapter, the philosophy within which measures for the benefit of repair, reconstruction or new development of the structure must be determined is discussed. The design philosophy matches the European norms, including Eurocode 0 (NEN-EN 1990 General), Eurocode 1 (NEN-EN 1991 Loads), Eurocode 7 (NEN-EN 1997 Geotechnology) and NEN 8700. The provisions for retaining structures have been included in chapter 9 of NEN 1997-1. In the Netherlands, the National appendix NEN-EN 1997-1/NB and the Supplementing provisions NEN 9097-1 apply and have been joined into one norm, Eurocode 7 [32], together with NEN-EN 1997-1 (general rules). This is collectively referred to as Eurocode 7 in this text. The national appendix contains a large number of sections from CUR 166 [7].

5.2   Reliability assessment

5.2.1   Reliability class

Reliability indexes for new build

The Eurocode distinguishes between reliability classes. In conformity with these design regulations, calculations for the reliability inspection can be made from a design with the systematics of characteristic load and strength parameters and partial reliability factors, the so-called load and material factors. The required reliability level in the EN norms deviates from the reliability level in the CUR handbooks [7] and [8] and the old Dutch NEN norms as also mentioned in section 4.3.3.

The underlying reliability requirements, in terms of minimally required reliability indexes (β-values) have been presented for new structures in table 14.

The highest reliability class in conformity with the Eurocode [17] concerns class RC3 with a β-value of 4.3 in regard to a reference period of 50 years. These reliability classes are described in the Eurocode as a great danger to the lives of dozens of people and large economic damage. For the testing of retaining structures in conformity with CUR 166 [7], it follows that CUR class III, with a β-value of 4.2 to 4.3 matches RC3.

Reliability indexes for existing structures

The values in table 14 apply to new structures. In case of existing structures, please refer to the Dutch Building Decree 2012 for testing, in conformity with the National Appendix of the Eurocode. For repair or reconstruction of existing structures, β-values that lie between the minimum values of existing structures and new structures apply.

In NEN 8700 [31] a minimum reference period of 15 years and a minimal reliability index of 1.8, 2.5 and 3.3 are prescribed respectively for testing the of existing structures with reliability class RC1, RC2 and RC3. For the repair of existing structures, reliability indexes of 2.8, 3.3 and 3.8 apply. The various values have been summarised in table 15.

Table 14 Safety classes for reference period of 50 years (new structures).

Risk of		EN 1990		CUR 162 + 166		NEN 6700	
economic damage	casualties	class	β_{50}	class	β_{50}	class	β_{50}
negligible	small				2.5 à 2.6		
negligible	small	RC1	3.3			1	3.2
small	considerable			II	3.4	2	3.4
considerable	high	RC2	3.8			3	3.6
very high	high	RC3	4.3	III	4.2 à 4.3		

Table 15 Safety classes for existing structures (review and reconstruction) and new structures.

NEN 8700 review existing structures		Repair/restoration existing structures		EN 1990 new structures	
class	β_{15}	class	β_{15}	class	β_{50}
RC1	1.8	RC1	2.8	RC1	3.3
RC2	2.5	RC2	3.3	RC2	3.8
RC3	3.3	RC3	3.8	RC3	4.3

For the repair of existing building structures, the new development level is the target level and the legally allowable deviation to the reconstruction level may only be employed in case of disproportional costs needed to meet the new development level [31]. Think of listed building structures or special structures, for instance. The implementation of monitoring the behaviour of the structure's critical parts must be taken into account as much as possible during repair.

For retaining walls on spread foundations or pile foundations, types 1, 2 and 3, the differentiation in reliability index occurs by differentiating load factors between the various reliability classes, according to the Eurocode. For this type of structure, the material factors for all reliability classes are equal. NEN 8700 [31] gives alternative load factors for existing structures, compared to the load factors applicable to new structures. The load factors from NEN 8700 are lower than those that apply to new structures. However, these load factors have been derived for buildings and bridges. The norm NEN 8707 is currently being developed for geotechnical structures. Therefore, no current load factors are available for the assessment of existing geotechnical structures.

Awaiting NEN 8707, it is advised to use the load factors derived from the assessment of existing buildings from NEN 8700 [31], where one must be aware of situations and loads that do not occur in the case of buildings. Think of soil pressure, for instance.

In these cases, one must act in accordance with the normative parts of the Eurocode, appendix B and C, including national appendixes. It is emphasised that, for repair, the new development level is the target level and a lower level may only be applied in specific cases. For sheet pile structures, quay walls type 4, the Eurocode states that differentiation in reliability index occurs by the differentiation of both material and load factors. It must be noted that this differentiation especially regards the soil parameters. NEN 8700 does not offer guidelines for a differentiation in material factors.

For this type of structure, however, the material factors from CUR 166 and the Eurocode may be used, which have been derived for a certain reliability index for a reference period of 50 years. This is a conservative length of time compared to a reference period of 15 years.

In this way, the materials factors that are also applied to new structures in CUR class 1, RC1 and RC2 respectively, (lowering of one reliability level) can be used for repair or reconstruction of existing structures with consequence class RC1, RC2 and RC3.

In conformity with the National Appendix of the Eurocode [7], structural elements may be categorised in a lower consequence class than the structure they are part of, on the condition that it can be expected that the consequences of failure are small.

5.2.2 Design approach

Three design approaches are mentioned in the Eurocode, from which a choice can be made. For every approach, a combination of sets of material factors (M), load factors (A), and factors on the total bearing capacity or resistance (R) is given, where the values of the factors are established by every country at its own discretion in the National Appendix.

- OB1.1 (combination 1): set of factors on the loads or the load effect (result of this load); material factors are equal to 1.
- OB1.2 (combination 2): set of factors on the material strength (material factors > 1); only load factor on unfavourable variable load.
- OB2: set with factors on the loads or on the load effect and on the resistance as a whole; material factors are equal to 1.
- OB3: set of factors on both the material strength and the loads.

In OB3, the calculation values of the various parameters are determined prior to the calculations. These parameters are determined by means of partial factors on regarding both the load and the materials and these are subsequently used for the calculations. This last method matches the semi-probabilistic reliability approach according to the Dutch Technical Principles for Building Structures 1990. In the Netherlands, OB3 is normative.

Although the Observational Method actually forms a separate design approach, this method is also often referred to as design approach OB4. In this design method, the design must be assessed during construction and adjusted if necessary. It is important that the following requirements are met.
1. Boundary conditions and allowable limits must be established.
2. An active monitoring must be performed during construction.
3. Management measures and mitigating measures must have been drawn up beforehand.
4. Close collaboration with an expert geotechnical advisor is required.

5.2.3 Reliability methods

Figure 5-1 shows a schematic overview from the Eurocode [17] of the methods available for calibration of partial factors (limit state) for the benefit of formulas for design and calculation and their mutual coherence. The probabilistic methods available for calibration of partial factors can be subdivided in:
- fully probabilistic methods (level III): is rarely applied to calibration for the design due to the lack of the required statistical data.
- first order reliability methods (FORM level II): this concerns the use of certain described approaches that can be considered sufficiently accurate for most structural applications.
- semi-probabilistic methods (level I): here, partial factors are applied to the various influence factors that have been established on the basis of probabilistic methods and that can deviate per country.

• deterministic method (level 0): this method works with characteristic values where no differentiation between the various influence factors occurs, but eventually, one overall reliability is calculated.

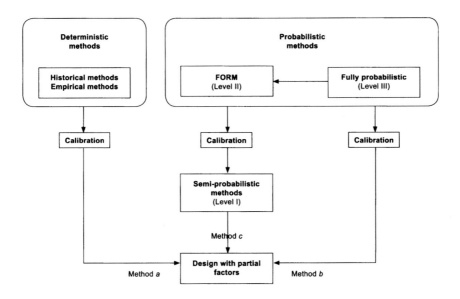

Figure 5-1 Schematic representation of reliability methods.

In the deterministic approach, the representative values, both on the load side and the resistance side (material side), are used. The calculated representative values are subsequently multiplied with one overall reliability factor, where no differentiation occurs between the various influence factors. This overall reliability factor can also be procured by dividing the representative resistance value with the representative load value. The upper half of figure 5-2 shows this deterministic approach.

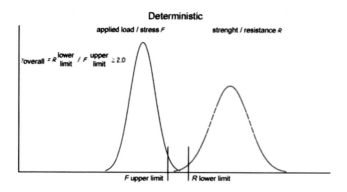

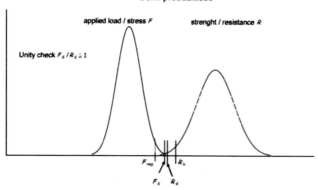

Figure 5-2 Deterministic and semi probabilistic approach.

In the Dutch norms and Eurocodes, a distinction is made into various influence factors that each display an individual spread. Both on the load side and the resistance side (material strength), it is therefore common to determine the calculation value for every load and material parameter individually. This is referred to as the semi-probabilistic approach, see the below half of figure 5-2. The semi-probabilistic approach forms the basis of the reliability philosophy in conformity with Eurocode 7 [32] and CUR 166 [7] and has been further elaborated by means of partial factors in the next section.

5.2.4 Influence parameters and partial factors

Load factors and material factors

In Eurocode 7 [32], a division is made into different types of structures. The partial load factors have been established based on the group in which the type of structure is categorised. A division is presented in table 16.

128

Table 16 shows the partial factors related to the various loads dependent upon the type of structure and the type of load. Structural loads and geotechnical loads (loads exercised on the soil) on structural elements present in the "geotechnical structure which is the foundation" such as retaining wall on piles, quay wall types 2 and 3, must be considered as loads from group B. The load factors must be determined in accordance with Eurocode 7, table A.3, column A1.

Retaining walls on spread foundations, quay walls type 1, possibly with nailing, have been defined as a "geotechnical structure which is not the foundation". Therefore, these are considered as ground retaining structures meaning that two different load calculations apply. The resting vertical loads and dead weights on the retaining wall (the ground segment right above the foot plate) must be categorised as group B, where the load factors must be determined in accordance with Eurocode 7, table A.3, column A1.

The loads that affect the wall via the soil pressure, are divided into group C. Also see table 16. This is also true for the negative friction which falls under "load soil", and therefore also group C. Here, the load factors must be determined in accordance with table A.3, column A2-Other.

For sheet pile walls under strain of bending, quay walls type 4, a division into group C applies to both structural and soil loads, where the load factors must be determined in accordance with Eurocode 7 [32], table A.3, column A2 Sheet pile wall. These loads concern all loads that affect the sheet pile wall via horizontal and vertical soil pressure.

To test the geotechnical vertical bearing capacity of sheet pile walls, both loads from the structure and those via the ground negative friction or vertical component of the active soil pressure must be divided into group B, also see table 16. The load factors according to table A.3 column A1 from Eurocode 7 apply to this.

129

Table 16 Categorisation structures in conformity with Eurocode 7.

No.[1]	Type of structure, type of calculation	Structural load	Soil load	Applicable type of quay wall
1	Spread foundations, concrete verification	group B	group B	type 1
2	Spread foundations, geotechnical bearing capacity verification	group B	group B	type 1
3	Pile foundation, moment plus normal force	group B	group B	type 2, 3
4	Pile foundation and sheet pile wall[3], geotechnical bearing capacity	group B	group B	type 2, 3, 4
5	Subsoil roof/wall structures	group B	group B	-
6	Slope stability/overall foundation stability	group C	group C	type 1, 2, 3, 4
7	Sheet pile wall under strain of bending	group C	group C	type 4
8	Gravity wall/L-wall/reinforced soil structure	group B	group C[2]	type 1

1) numbers 1 to 5 concern geotechnical structures which are foundations, numbers 6 to 8 concern geotechnical structures which are not foundations.

2) loads that affect the wall via the soil pressure must be categorised in group C.

3) to test the geotechnical vertical bearing capacity of sheet pile walls, both loads from the structure and those via the soil (negative friction or vertical component of active soil pressure) must be categorised in group B.

The load factors from Eurocode 7 were originally based on RC2, where the factors for RC1 and RC3 are derived with K factors from Eurocode 7, table B3. Here, both the permanent and the changeable loads have been multiplied with a factor 0.9 for RC1 and table 17 summarises the partial load factors for the various reliability classes RC1, RC2 and RC3 (new structure).

A partial factor on volume weight $\gamma G;w = 1.0$ applies to the (differences in) water pressure on the retaining structure, where the normative calculation values of the water levels on both sides of the retaining structure must be calculated. If a retaining structure exclusively or practically exclusively retains water, a geotechnical load factor $\gamma F = 1.2$ must be applied for the ultimate limit state. The results calculated in the usability limit state must be multiplied with a factor 1.2 in accordance with step 6.5 x 1.2 from CUR 166. A distinction between sheet pile walls and gravity/retaining walls on spread foundations (which are not foundations) is also made for the material factors of the strength parameters. In case of the latter retaining walls, no distinction is made between the reliability classes. To calculate the calculation value of the loads on a retaining wall on spread foundations, the calculation values must have been applied to strength parameters in the active soil pressure, a partial material factor of 1.0 applies to the volume weight of the soil on the load side.

Table 17 Partial load factors retaining structures – new structure (Eurocode 7).

Application	Application	Symbol	Value RC1	Value RC2	Value RC3
Group B (Column A1)	Permanent load, unfavourable[1]	y_G	1.2	1.35	1.5
	Permanent load, unfavourable[2]	$y_{G \times \xi}$	1.1	1.2	1.3
	Permanent load, favourable	$y_{G;stb}$	0.9	0.9	0.9
	Changeable load, unfavourable	$y_{Q;dst}$	1.35	1.5	1.65
	Changeable load, favourable	$y_{Q;dst}$	0	0	0
Group C (Column A2-sheet pile wall)	Permanent load, unfavourable	y_G	1.0	1.0	1.0
	Permanent load, favourable	$y_{G;stb}$	1.0	1.0	1.0
	Changeable load, unfavourable	$y_{Q;dst}$	1.0	1.1	1.25
	Changeable load, favourable	$y_{Q;dst}$	0	0	0
Group C (Column A2-other)	Permanent load, unfavourable	y_G	1.0	1.0	1.0
	Permanent load, favourable	$y_{G;stb}$	1.0	1.0	1.0
	Changeable load, unfavourable	$y_{G;dst}$	1.2	1.3	1.4
	Changeable load, favourable	$y_{G;dst}$	0	0	0

1) only applies to small changeable loads (Q/G < 0.2)
2) only when share changeable load is large (in conformity with NEN-EN 1990/NB apply ξ = 0.89)

The soil pressures on the wall must be considered as geotechnical loads, also when these are partly the result of, for instance, a changeable load on ground level. In appendix 2, 3 and 4, the various partial factors per type of retaining wall have been presented, for new structures, for repair or reconstruction and for the testing of existing structures (disqualification level) respectively.

Geometric factors

In addition to the abovementioned partial load material and resistance factors, influence factors also apply in conformity with Eurocode 7 [32] that are related to the unreliability of the geometrics for excavation level (retaining height of the wall) and the water levels (drop across the wall length).

The calculation value of the retaining height is calculated by applying an excavation margin to the passive side for an unchanged ground level on the high active side. When choosing the excavation margin, the extent to which the excavation is monitored on the construction site must be taken into account. In case of normal monitoring, the following applies:

$$\Delta a_{excavation} = 0{,}1 \cdot h \ \forall \ \Delta a \leq 0{,}5 \text{ m} \qquad [-]$$

in which:

$\Delta a_{\text{excavation}}$ = excavation margin [m]

h = Height equal to the minimal wall height above excavation level to top retaining wall (for retaining wall type 4 with anchorage, the difference between excavation level and lower anchorage level applies[m]

When it has been established that the excavation level has been reliably monitored, $\Delta a = 0$ can be assumed in conformity with Eurocode 7. In case of major uncertainties, such as during wet excavations of the bottom for quay structures, higher values could apply, in which dredging tolerances and propeller action must be taken into account, see CUR 211 [8]. The excavation margin in Eurocode 7 is not dependent upon the reliability class (RC).

For the drop across the wall length, the calculation values of the most unfavourable water levels must be applied in the ultimate limit state. These are dependent upon the situation in which these are tested. In the calculation, the margins must be applied in conformity with table 18. For the calculation value of the (ground)water level and the water pressure at the high side, the top side of the retaining wall may be used as the upper limit if a higher water level is not physically possible.

Uncertainty factors in the location of soil investigation

Additionally, uncertainties with regard to the location of cone penetration tests and the extent of these also play a role in the shaft friction of piles and grout bodies of anchors. For this purpose, correlation factors have been established in the mathematical model for the calculation of the pressure and tensile bearing capacity in conformity with Eurocode 7 [32]. These are dependent upon the number of tests with which the characteristic value can be calculated from the calculated value of the shaft friction. The values have been presented in table A.10 of Appendix A of Eurocode 7.

Table 18 Partial factors and surcharges water levels.

Parameter	Value RCO[3]		Value RC1		Value RC2		Value RC3		Calculation value
	γ^0 [-]	Δa [m]	γ^0 [-]	Δa [m]	γ^0 [-]	Δa [m]	γ^0 [-]	Δa [m]	
Phreatic level low side	1.30	0.15	1.70	0.20	1.90	0.25	2.10	0.25	$\max_{(\mu + \gamma \cdot \sigma \,;\, \mu + \Delta a)}$ [2]
Phreatic level high side	0.66	0.05	0.87	0.05	1.18	0.05	1.50	0.05	$\max_{(\mu + \gamma \cdot \sigma \,;\, \mu + \Delta a)}$

1) if the probability distribution is known (μ and σ).
2) and/or minus($\mu - \gamma \cdot \sigma$; $\mu - \Delta a$).
3) assuming CUR class I, matching $\beta = 2.5$.

5.3 Structural calculation

5.3.1 Calculation steps

In general

In the next sections, the calculation steps for the geotechnical and structural design of the quay wall have been presented as these apply in conformity with Eurocode 7, and have been translated to the specific types of quay walls as these have been defined in the previous chapters.

Calculation steps retaining wall on spread foundations (type 1)

For the calculation of a retaining wall on spread foundations, the failure mechanisms stated in section 4.3.1 must be considered. Based on this, a step plan is presented below.

The calculation steps to be followed are:

1. Calculation loads normative situations:
    a. determination load situations, phasing and normative construction phases, and cross sectional profiles;
    b. determination surcharges, loads due to soil pressure, water and external loads.

2. Calculation and testing vertical bearing capacity:
    a. determination load and effective dimensions foundation surface;
    b. determination influence depth;
    c. testing vertical bearing capacity in undrained state (if applicable);
    d. testing vertical bearing capacity in drained state;
    e. testing on puncturing (if applicable);

133

    f. testing on squeezing (if applicable).

3. Calculation and testing resistance against horizontal sliding:
    a. determination load and effective dimensions foundation surface;
    b. testing horizontal bearing capacity in undrained state (if applicable);
    c. testing horizontal bearing capacity in drained state.

4. Calculation and testing general stability.

5. Calculation and testing tilt stability (forces with high eccentricity).

6. Calculation and testing failure of the structure:
    a. dimensioning structure retaining wall, testing strength;
    b. testing failure structure due to displacement of the foundation (ULS).

7. Calculation and testing settlement and settlement differences:
    a. determining load and effective dimensions foundation surface;
    b. calculation tension spread in the depth;
    c. calculation settlement;
    d. calculation settlement difference;
    e. testing on deformation requirements.

8. Testing hydraulic failure mechanisms.

The spread foundation must normally be calculated in conformity with chapter 6 of Eurocode 7 and informative appendixes:
- Appendix C: mobilisation of soil pressure for horizontal bearing capacity.
- Appendix E: semi-empirical method for calculation bearing capacity from pressure meter tests.
- Appendix G: calculation bearing capacity spread foundations and rock foundations.
- Appendix H: limit values for deformations in the usability limit state.
- Appendix J: reference list monitoring.

Calculation steps retaining wall on pile foundation (types 2 and 3)

For the calculation and testing of tension loaded foundation piles, the failure mechanisms stated in section 5.2.1.2 must be considered. This can be done by means of the step scheme:

The calculation steps to be followed are:

1. Calculation loads decisive situations:
   a. determination load situations, phasing and decisive construction phases, and cross sectional profiles;
   b. determination surcharge, loads due to soil pressure, water and external loads.

2. Calculation and testing vertical bearing capacity (pressure):
   a. determination pile head load;
   b. choice of type of pile head with specific properties;
   c. choice of construction order and influences on cone distance (excavation, OCR, construction);
   d. determination negative and positive friction zones and calculation negative friction;
   e. calculation point bearing capacity per cone penetration test;
   f. calculation shaft friction over positive friction zone per cone penetration test;
   g. calculation net decisive pile bearing capacity;
   h. testing vertical bearing capacity.

3. Calculation and testing vertical bearing capacity (tension):
   a. determination (external) pile head load (pressure and tension) for tension piles;
   b. choice of type of pile with specific properties;
   c. choice of construction order and influence on cone resistance (excavation, OCR, construction);
   d. determination tension zone;
   e. calculation value cone resistance;
   f. calculation effect of pile installation (factor f1);
   g. calculation effect of application tension load (factor f2);
   h. calculation of the bearing capacity;
   i. calculation soil weight on pile after removal;
   j. testing tension bearing capacity.

4. Calculation and testing laterally loaded piles (horizontally loaded on pile head):
   a. determination soil parameters (soil pressure factors, horizontal soil compressure factors);
   b. determination possible boundary conditions for pile head (translation and rotational spring);
   c. determination flexural rigidity piles (top metre as torn in case of fixation);
   d. determination parameters in case of interaction with pile groups;
   e. determination axial spring stiffness;
   f. calculation moment, transverse forces and displacements.

5. Calculation and testing of general stability.

6. Calculation and testing failure of the structure:
   a. dimensioning structure retaining wall, concrete floor;

135

b. testing strength requirements of tension pile (ULS);

c. testing crack width of tension pile (SLS);

d. testing failure of structure due to displacement of the foundation (ULS).

7. Calculation and testing settlement and settlement differences:
    a. determination pile head load, pile stiffness and type of settlement curve;
    b. calculation settlement from point and elastic settlement;
    c. calculation compression cohesive layers underneath pile group;
    d. calculation primary and secondary settlement;
    e. calculation settlement difference;
    f. calculation rise tension piles from mobilisation shaft and elastic lengthening;
    g. calculation rise from contribution from lifting soil layers;
    h. testing deformation requirements (SLS).

8. Testing hydraulic failure mechanisms.

Steps 3f, 3g and 3i only apply to the tension loaded pile groups and do not apply to single tension piles. These can be omitted for single piles.

Step 3 only relates to piles that are horizontally loaded. In the case of the presence of anchorage, nailing or soil reinforcement, additional calculation steps apply. For anchorage, please refer to the calculation steps concerned for quay wall type 4 (sheet pile wall) in this section.

The piles must be calculated and tested in conformity with chapter 7 of Eurocode 7. Additionally, the appendixes of NEN=EN-9997-1 provide further information regarding pile foundations (informative):

• Appendix H: limit values for deformations in the usability limit state.

• Appendix J: Reference list monitoring.

Calculation steps sheet pile wall as retaining wall (type 4)

The criteria for the calculations and the reliability requirements for the benefit of the calculation of sheet pile structures have been established in chapter 9 of Eurocode 7. These have partly been taken from the guidelines for sheet pile structures, as described in CUR 166. This includes a step plan for testing, assuming a mathematical model according to the spring supported beam. The various steps from CUR 166 that have been included in Eurocode 7 are presented on the following pages (articles):

1. Determination of decisive principles (art. 9.7.1(a)):
    a. determination of reliability class, decisive cross sectional profiles and construction phases;
    b. determination of loads, water pressures, water over-tension, surcharges, lifespan.

2. Determination of characteristic values of the parameters (art. 9.7.1(b)):

a. determination of soil parameters;
b. determination of sheet pile parameters, including possible corrosion allowance.

3. Determination of calculation values of the parameters (art. 9.7.1(c) up to (g)).

4. Choosing calculation scheme A (calculation value in all phases) or B (calculation value only in the phase to be tested (art. 9.7.1(h)).

5. Calculation minimal embedding depth (art. 9.7.1(i) up to (j)):
   a. calculation embedding by means of sheet pile wall calculation;
   b. testing vertical bearing capacity (art. 9.7.5);
   c. testing functional requirements (for instance water sealing).

6. Dimensioning calculations with variation of high and low soil compression factors and high and low groundwater level (art. 9.7.1(k)):
   a. testing ULS + k and high GWS on low side (dimensioning calculation No. 1);
   b. testing ULS + k and high GWS on low side (dimensioning calculation No. 2);
   c. testing ULS + k and low GWS on low side (dimensioning calculation No. 3);
   d. testing ULS + k and low GWS on low side (dimensioning calculation No. 4);
   e. testing SLS + k with partial factors at 1.0 (dimensioning calculation No. 5);
   f. testing SLS + k with multiplication results with 1.2 (supplementing calculation).

7. Testing of capacity moment sheet pile wall (art. 9.7.1(l)):
   a. calculation of maximum moment M;
   b. calculation of capacity moment on the basis of elasticity or plasticity;
   c. determination of possible reduction factor resulting from large differences in water pressure.

8. Testing strength sheet pile wall on normal forces and transverse forces (art. 9.7.1 (n)):
   a. determination situations with maximum transverse forces and normal forces;
   b. testing strength sheet pile wall by means of material-bound Eurocodes, also see below in elaboration step 8.

9. Dimensioning anchorage and stamping (art. 9.7.1

(o) up to (r)):
   a. calculation possibly with 5% upper limit of the anchor/stamping stiffness force;
   b. determination normative anchor/stamping force;
   c. determination calculation value of the normative and representative anchor/stamping force;
   d. testing strength and tension bearing capacity including failure, corrosion, improper use, test load, temperature influences.

10. Deformation test (art. 9.7.1 (s)).

11. Testing other mechanisms (art. 9.7.1 (t) up to (v)):
    a. testing Kranz stability (if applicable);
    b. testing general stability.

c. testing soil breach (exceeding bearing capacity subsurface);
d. testing failure due to washing away;
e. testing conductibility installation sheet pile wall;
f. testing vibration and/or sound influences on the surroundings.

12. Verification sheet pile wall calculation.

The retaining structure must be calculated in conformity with chapter 9 of Eurocode 7 [32]. Additionally, the appendixes contain additional information regarding retaining structures (informative):

- Appendix C: Example procedures for the determination of the limit values of the soil pressure.
- Appendix H: Limit values for deformations in the usability limit state.
- Appendix J: Reference list monitoring.

Section 5.6.2 elaborates on the calculation of a sheet pile structure as retaining wall by means of a calculation example.

5.3.2   Mathematical models

Proper inspection and testing of the failure mechanism requires a proper mathematical model, this being a model that simulates the actual situation as accurately as possible. However, an ideal mathematical model does not exist: a model contains uncertainties that can be discounted with reliability factors in the testing of the design. Mathematical models aim to describe the essential behaviour of a structure under certain circumstances, due to which it can be inspected whether the design and testing requirements are met.

Models can vary from simple (empirical) relationships to highly advanced mathematical models, such as the Blum method versus FEM analyses for retaining structures. Material behaviour can also be considered as being purely elastic to being complexly elastoplastic.

Although a simple model requires little input parameters, the results will usually be on the cautious side of reality, while for complex models, the uncertainty largely consists of the sum of all uncertainties per parameter. In the first case, the model uncertainty is relatively large, while the second model is expected to be more accurate and has a model uncertainty with a lower coefficient variation.

5.3.3   Behaviour of the current structure

If an existing structure is repaired, renovated or rebuilt, information on the current (disqualified) structure is vital. Several examples of this are:

- The presence of old piles that could hinder the installation of new elements, such as piles, anchors, sheet piles, etc.

- If the current structure is water-permeable, this can significantly influence the water levels behind the structure. Repair of the water permeability could reduce the washing away of soil and water, but could also lead to higher loads on the structure.
- Presence of (parts of) old quay structures that have lost their function.
- Old basalt slopes or deposits that could hinder the installation of piles, anchors and sheet walls, etc.

5.4 New structures and repair of listed buildings

5.4.1 Introduction

When assessing and altering listed buildings, a balance must be found between the opportunities for conservation of the current and of the future function and the possibilities for the conservation of the special character and specific value of the listed building.

For a listed building and all accompanying parts, conservation is nearly always preferable. Repair of parts during the building's lifetime will, however, sometimes be necessary. Before action is taken, the wishes of the user must be considered in comparison to the opportunities which the listed building provides.

In the first place, it must be assessed how adaptations to the front side, for which the functional requirements have been established, can lead to renewed approval of the structure based on the adjusted requirements, as described in the test scheme. If repair of certain parts is inevitable, the repair must be carried out as sustainably and responsibly as possible, and any building activities should be conducted in an appropriate style.

5.5 Effects of renovations and new structures

In addition to materials and the aforementioned test criteria for renovation of existing structures, the effects of renovations or new construction of listed building structures must also be taken into account during the construction. Examples include:
- effect of vibrations during work activities;
- displacements (caused by vibrations as well as heavy loads/excavations during the construction phases);
- noise and construction nuisance.

Vibrations during building activities or repair activities can have various effects. First, vibrations can influence the structural damage of the structure. In order to adequately manage the degree of vibrations, the vibration intensity should be tested with the limit values that can be determined with the guideline SBR-A [34]. This includes the limit values for vibrations with regard to settlement of the foundation. Guideline SBR-B [35] includes the limit value for vibrations with regard to nuisance and noise.

Guideline SBS-A is based on limit values at which the risk of new damage occurring is kept to a minimum. Exceeding the limit value will not necessarily lead to damage. There is only risk of actual damage once the limit value is exceeded to a certain degree. If the limit value is not exceeded, the chance of damage is relatively small. That is to say, less than 1%. The visibility of existing, camouflaged cracks (e.g. behind the new stucco) is not taken into account. This should therefore be regarded as an acceptable risk.

SBR-A makes a distinction between settlement-sensitive foundations and settlement-insensitive foundations. In parallel, building categories were determined from category 1 to category 3 with decreasing structural condition. The highest category applies to listed building, namely category 3. The influence of the vibrations with regard to the construction method of the existing structure as well as the influence of the vibrations with regard to a possible settlement-sensitive foundations are examined in the vibrations analysis.

The SBR-B vibration guideline includes guidelines for nuisance. Unlike in the case of damage, not the top value of the vibration speed but the effective value $v_{max;eff}$ and the vibration strength over a certain assessment period is important for nuisance. The guideline recommends to timely inform the respective residents or users about the nature of the vibrations and to involve them in the planning of the work activities.

The method of construction can be determined based on the limit values for vibrations. On the basis of a vibration prediction, one can determine how pile-driving activities should be performed. Boundary conditions can be established for type vibrator/ram, dominant frequency or one may decide to perform the activities in a vibration low/free manner. In regard to the target values for nuisance, the time of day (afternoon) and duration (spread across the day) of the work activities can be determined, depending on the building and housing function in the environment in which nuisance occurs. The effects can be managed by monitoring the vibrations according to the aforementioned guidelines.

5.6 Practical example

To provide insight into the way in which the safety philosophy and design steps are applied in practice, the following pages include a calculation example of the new construction of a quay structure, assuming quay wall type 4.

5.6.1 Reconstruction Conradkade in The Hague along the tram line

The project site is located along the Conradkade in The Hague. This case will focus on a certain part of the project, namely the location of water section 50632. This section has a length of approx. 150m.

Figure 5-3 water section 50632, Conradkade (Source: www.bing.com/maps).

In the past, the tram line served as a railroad to transport coal to the power station nearby. Later, the railway was used by the Haagse Tramweg Maatschappij (HTM; The Hague Tram Company) With the arrival of Randstadrail, the tram line was to be made suitable for Randstadrail vehicles. This new material is heavier than the material of HTM, as a result, the load on the canal bank increases. In the previous situation, there was a bank with a steep slope and anchored wooden sheet piling as a toe construction. However, this was not suitable for supporting the heavier loads and was therefore replaced with a construction consisting of anchored steel sheet piling.

The steel sheet pile structure was designed according to the guidelines of CUR 166. In this case, the design of the new structure will be explained, in particular addressing the parts of the design that specifically relate to urban quay walls. Think of environmental influence and the influence of an old - to be replaced - quay structure.

A typical cross section of the previous situation is shown in the figure below.

141

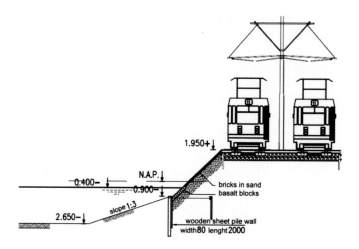

Figure 5-4 Principle cross section old situation.

In the old situation, the quay structure consisted of an anchored wooden sheet pile wall with a slope of approximately 1:1 (d:h) The slope was covered with a stone setting. Because no sounding of the underwater slope and bottom level was available, a flood prevention profile was used for this. This flood prevention profile is shown in figure 5-4. In the old situation, the distance between the tram line and the wooden sheet pile wall was approx. 2.5 m.

Design new structure

For the design of the new structure, a soil analysis and design calculations were performed. Several aspects of this will be discussed here briefly.

For the design of water section 50632 three cone penetration tests and one drilling test were performed right along the existing tram line. The soil structure was determined on the basis of an interpretation of the soil analysis. This is presented in table 19.

Table 19 Representative soil description at the Conradkade location.

Top layer [m, *NAP*]	Soil layer
+1.9 to +2.7	Sand, low density
-1.0	Bog, soft
-1.5	Sand, low density
-5.0	Sand, medium density
-6.5	Clay, sandy
-6.7	Sand, medium density
-8.3	Clay, sandy
-8.9	Sand, low density
-10.2	Clay, sandy
-10.5	Sand, high density

An important boundary condition of the project was that the flow profile of the waterway could not be scaled down. Therefore, the new steel sheet pile wall had to be installed behind the existing wooden sheet pile wall. To install the new steel sheet pile wall behind the wooden sheet pile wall, the anchorage of the existing structure had to be removed. To prevent instability during construction, an auxiliary sheet pile wall was developed, which was then placed in front of the existing sheet pile wall. The area between the auxiliary and the wooden sheet pile wall was filled with sand, so that the wooden sheet pile wall could be supported after cutting through the anchorage.

The sheet pile wall calculations were conducted according to the guidelines of CUR 166. Following these guidelines, sheet pile profile AZ18 was assigned to both the free standing and anchored definitive sheet pile wall. A schematic representation of the two sheet pile walls is shown in figure 5-5.

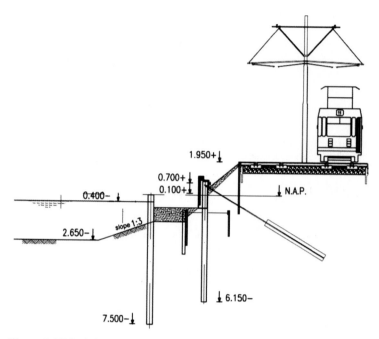

Figure 5-5 Principle cross section definitive and auxiliary sheet pile wall.

In this case, special attention had to be paid to the heavy loads (resulting from the tram) on the sheet pile structure. In addition to the vertical loads from the weight of the tram, horizontal loads as a result of wind pressure and lateral thrusts on the tram's movement during transport also had to be taken into account.

Furthermore, pipe intersections required special attention. In this case, a single gas pipe was present in a single location, which intersected with both the wooden sheet pile wall and the new steel sheet pile wall. Therefore the sheet pile wall couldn't be placed as deep in this location. As a result, the vertical load had to be carried by the adjacent sheet pile wall instead. To guarantee the vertical balance of the adjacent sheet pile walls, these were installed approx. 2 metres deeper than the other piles. The connection between the pipe and sheet pile wall was designed in such a way that the sand behind the pile cannot wash out.

Objects that may be damaged due to the vibrations during the work activities are located a short distance from the to be constructed sheet pile walls. This concerns the existing tram line at an approx. 7.5 metres distance (the nearest tram line is 2.5 metres away, but was temporarily taken out of service during the work activities) and the structures at an approx. 15 metres distance. What was especially important to take into account for the tram line, was the compaction of the sand layer and the resulting setting of the tram line. For the structures, the occurrence of direct vibration damage was of concern. The vibration risk analysis showed that the risk of damage to adjacent structures would be minimal. However, the vibrations and deformations had to be monitored during execution of the work.

Construction problems

During the installation of the definitive sheet pile wall, the slope became unstable, see figure 5-6. The work was immediately suspended.

Figure 5-6 Instability during construction.

An analysis of the instability showed that the design had taken into account the effect of vibrations on the adjacent structures and tram line, but not the stability of the slope. During construction, the stone setting on the slope was removed, causing the stability of the slope to decrease significantly. The slope without stone setting in the old situation was hardly sufficient, and as a result of vibrations during installation of the sheet pile wall along the slope, it became unstable.

Following this instability, a comprehensive survey of the underwater profile was conducted. This showed that the waterway was 0.5 m deeper than in the flood prevention profile. It also showed that, according to the flood prevention profile, the underwater support bank against the wooden sheet pile wall was entirely washed away. In response, the underwater profile was complemented up to the flood prevention level by applying crushed sand to the bottom of the waterway.

To prevent further instability, the slope behind the auxiliary sheet pile wall was complemented and a gentle slope from the head of the auxiliary sheet pile wall was installed at the level of the tramline. In this way, the bank was stabilised, so that vibrations caused by the work activities involving the definitive sheet pile wall would no longer caused instability. As a result, no further problems arose during construction.

Conclusions

It is possible to draw several important conclusions from this case.

One should accurately be aware of the geometrics of the old situation, both along the high and low side of sheet pile wall. This also applies to the (often more difficult to measure) underwater geometrics for the sheet pile wall. However, one must always remember that the actual geometrics can vary widely from the old drawings or flood prevention profiles. The stability during the construction phase must carefully be examined. The phases of the work activities have a crucial influence on the construction stability. The boundary conditions for the construction must be clear to all parties. The adjusting phasing is shown in the diagram below.

Phase 1:
1. Tramline along the slope is no longer in use.
2. Installation of auxiliary sheet pile wall 2 metres in front of the wooden sheet pile wall.

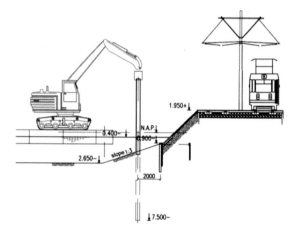

Figure 5-7 Schematic representation phase 1.

Phase 2:
1. Filling of area between auxiliary sheet pile and wooden sheet pile (black).
2. Cutting existing anchors.
3. Removal of clinkers and basalt.
4. Supplementing area behind sheet pile wall (black-white).
5. Installation of definitive sheet pile wall.

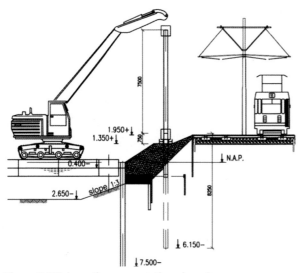

Figure 5-8 Schematic representation phase 2.

Phase 3:

1. Excavation for definitive sheet pile wall.
2. Placement of anchors.

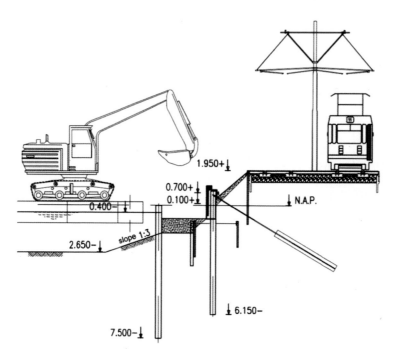

Figure 5-9 Schematic representation phase 3.

Phase 4:

1. Supplementing soil behind sheet pile wall.
2. Covering slope again.
3. Removal of auxiliary sheet pile wall.
4. Restoring soil profile in front of sheet pile.
5. Reinstating of tram line.

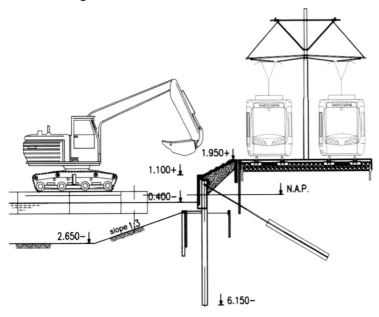

Figure 5-10 Schematic representation phase 4.

5.6.2   Reconstruction Rijnkade in Arnhem

In the context of the reconstruction of the Rijnkade in Arnhem, the old low quay wall was replaced by a 700 anchored steel sheet pile wall that was placed in front of the old quay wall. In this example, the sheet pile wall is dimensioned in accordance with the step plan and the safe pile-driving requirements in conformity with sections 5.2 and 5.3.

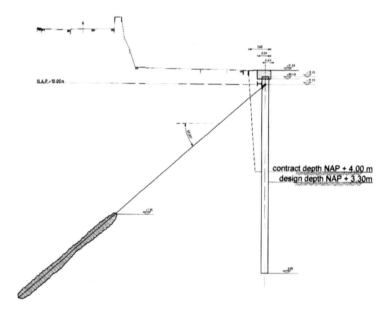

contract depth NAP + 4.00 m
design depth NAP + 3.30m

Figure 5-11 New sheet pile wall situation.

Figure 5-12a and b View new sheet pile wall (above) and sounding S09 (right page).

150

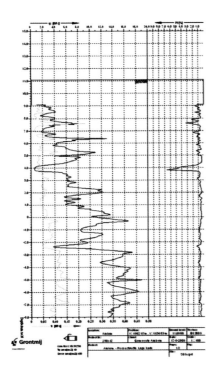

Step 1a: Determination reliability class, cross section and construction phases.

Given the fact that the current Low Quay lies within the influence zone of the High Quay (which acts as primary flood prevention), the new sheet pile structure should be designed in accordance with reliability class RC3 (a lower reliability class on the basis of RC1 applies for the construction phase). A β-value of 4.3 applies to RC3. As a result of the inclusion of the length-effect, the value in accordance with CUR 166 was increased to 4.9. Finally, a longer lifetime than the reference lifetime of 50 years is also required. For a lifetime of 100 years and in accordance with paragraph 2.4.7 of CUR 166, the β-value was further increased (by 0.2) to a final 5.1. On this basis, the different partial factors were scaled up.

Examined is cross section DWP5. The design phases are considered to be: the building phase with sheet pile wall and supplement and the final situation with minimum soil profile (construction depth $NAP + 3.3$m) and maximum water level drop (GWS at $NAP + 8.45$ metres and open water level MLW at $NAP + 7.05$m).

Step 1b: Determination loads, water pressures, excess water pressure, surcharges

A load from construction vehicles with a uniformly distributed load of 10 kPa has been determined for the building phase. For the final phase, a uniformly distributed load of 30 kPa and the aforementioned water level difference has been determined. Bollard load and mooring loads have not been taken into account in the example calculations.

Step 2a: Determination characteristic values of the soil parameters.

On the basis of cone penetration test S09, see figure, soil parameters have been determined, which have been assessed on the basis of table 2.b of Eurocode 7. The values are summarised per soil layer in table 20.

On the basis of the above soil parameters, the active and passive soil pressures have been determined in D-Sheet Piling. Three load branches were used for the subgrade reaction modules, of which only the value of the first branch is listen in Table 20 and a single relief branch.

Table 20 Characteristic values geotechnical parameters

Soil type	Upper side layer [m+NAP]	γ_d [kN/m³]	γ_n [kN/m³]	c' [kPa]	Φ' [°]	δ' [°]	$k_{h;1}$ [kN/m²]
Sand, clayey (supplement)	+11.0	17	19	0	27.5	18.3	12.000
Sand, low to medium density	+6.5	18	20	0	30.0	20.0	12.000
Clay, medium to very sandy	+4.2	17	17	0	27.5	18.3	6.000
Sand, low to medium density	+3.8	18	20	0	30.0	20.0	12.000
Sand, medium density	+0.5	18	20	0	32.5	21.7	20.000

Step 2b: Determination of sheet pile parameters, including additional corrosion

Profile AZ37-700 was initially used to measure the sheet pile wall. In addition, corrosion reduction was calculated for a lifetime of 100 years. This example is based on a wall thickness of 8.6mm in the atmospheric zone, the splash zone and the low-tide zone up to *NAP* +7 metres and a wall thickness of 5.0 mm in the underwater zone and soil zone beneath it. The relatively high values are a result of an aggressive environment (port, industry, existing sewage), previously contaminated Rhine water and additional oxygen influences as a result of the helical movements of ships.

The sheet piling parameters are as follows:
- steel sheet pile wall AZ37-700 with steel quality S355 GP;

- Head level: NAP +11.00 m;
- Anchor level: NAP +10.00 m (on the basis of statistical analysis water levels);
- wall thickness body: t = 12.2 mm;
- wall thickness flange: t = 17.0 mm;
- flange width: b = 2 x 225 = 450 mm;
- paint surface: A = 1.46 m^2 /m^2;
- point surface: A = 0.023 m^2 /m;
- mass: m_0 = 176 kg/m;
- flexural rigidity: EI_0 = 194.000 kNm^2 /m;
- moment of resistance: $W_{el;0}$ = 3.703 cm^3 /m;
- Reduction oblique bending: none.

For the sheet pile wall above NAP +7.0 m:
- corrosion reduction 8.6 mm: 0.50;
- flexural rigidity: EI_{cor} = 102.204 kNm^2 /m;
- moment of resistance: $W_{el;cor}$ = 1.951 cm^3 /m.

For the sheet pile wall below NAP +7.0 m:
- corrosion reduction 5.0 mm: 0.72;
- flexural rigidity: EI_{cor} = 140.647 kNm^2 /m;
- moment of resistance: $W_{el;cor}$ = 2.684 cm^3 /m.

Table 21 Partial material factors for retaining structures.

Parameter	Symbol	Value RC3	a_i [-]	V_i [-]	$y_{M;corr}$ [-]	Value RC3$_{corr}$
Angle of internal friction	$\gamma_{\varphi'}$ [1)	1.2	0.8	0.10	1.066	1.3
Effective cohesion	γ c'	1.4	0.8	0.20	1.135	1.6
Soil compression factors, E-moduli, unfavourable	γ_{kh}, γ_E	1.0	-	-	-	1.0
Soil compression factors, E-moduli, favourable	γ_{kh}, γ_E	1.3	0.8	0.10	1.066	1.4
Changeable load, unfavourable	$\gamma_{Q,dst}$	1.25	0.8	0.10	1.066	1.35
Changeable load, unfavourable[2)]	$\gamma_{Q,dst}$	1.65	0.8	0.10	1.066	1.75
Permanent load, unfavourable[2)]	$\gamma_{G,dst}$	1.5	0.8	0.05	1.032	1.55

1) factor is applicable to tan φ'.

2) applicable to group B (assess vertical balance).

Step 3: Determination calculation value of the parameters

The calculation values of the soil parameters have been determined on the basis of the partial factors from table 21. For the final situation, the factors have been corrected from $\beta = 4.3$ to $\beta = 5.1$. A correction for a higher reliability index also applies to the strength of the steel sheet pile wall, see test step 7. For steel structures, a variation coefficient of $V_i = 0.10$ and an influence factor of $\alpha_i = 0.8$ can be assumed in conformity with CUR 166. The following equation follows from this:

$$y_{m;corr} = e^{0,8 \cdot (5,1-4,3) \cdot \sqrt{\ln(1+0,12)}} = 1,066$$

The corrected values of the soil parameters and the permanent and changeable loads have been summarised in table 21.

With regard to the margins in groundwater level, the following applies:
- ground side: NAP +8.45 metres + 0.05 metres = NAP +8.50 m
- water side: NAP +7.05 metres − 0.25 metres = NAP +6.80 m

A value of $\Delta a_{excavation} = 0.1 \cdot (10.0 - 4.0) = 0.60$ metres and where $\Delta a_{excavation} \leq 0.50$ m follows for the margin in excavation level, in conformity with Eurocode 7, on the basis of a ground level (contract depth) at $NAP + 4.0$ metres and an anchor level at $NAP + 10.0$ metres. Higher values may apply to quay structures. In conformity with table 5.1 from CUR 211 [8], a total excavation margin of 0.7 metres below contract depth follows, leading to a structural depth of $NAP + 3.3$ m. Furthermore, the design includes a disturbance level of 0.35 metres, which means that $NAP + 2.95$ m has been used for the ultimate limit state in the final situation.

Step 4: Choose mathematical scheme A or B

As two different reliability levels apply to the two phases, calculations have been conducted in accordance with calculation scheme B:
- phase 1: RC1 (construction phase after placement of sheet pile wall and anchors raising to $NAP + 11.0$ m).
- phase 2: $RC3_{corr}$ (end phase with maximum water drop and loads and deepest bottom level).

Step 5a: Calculate fixation by means of a sheet pile wall calculation

Table 22 includes the results of the iterative calculation for the required sheet pile wall fixation, as conducted with computer programme D-Sheet Piling. For a point level above NAP -2.0 metres, the sheet pile wall is instable. Therefore, a point level at NAP -3.0 m was initially chosen.

Step 5b: Test vertical bearing capacity

154

The anchor force from table 22 has been calculated on the basis of an anchor angle of 40° with the horizontal. As result, the sheet pile wall is vertically loaded with an anchor force. In order to calculate the vertical load, the sheet pile wall must be loaded on the basis of the load factors from table 21 as a foundation element in accordance with group B (see table 6.2.2).

The calculation can be performed by means of an extra calculation phase based on the SLS approach, times an overall factor on the calculated loads. The following applies:
- applying load factor to changeable load $\gamma_{Q;dst;corr}$ = 1.75/1.55 = 1.15.
- multiplication calculated vertical loads (permanent) with $\gamma_{G;dst;corr}$ = 1.55.

The following follows from the calculation:
- anchor force:
  $F_{A;max}$ = 192.4 kN/m.

- vertical load active soil pressure:
  $F_{v;active}$ = 276 kN/m;

- vertical load passive soil pressure:
  $F_{v;passive}$ = -286 kN/m;

- vertical anchor force:
  $F_{v;A;max}$ = 192.4 • sin 40° = 125 kN/m;

- total vertical load on sheet pile wall point:
  N_{Ed} = 1.55 • (276 - 286 + 125) = 178 kN/m.

On the basis of the fixation length, the following follows, with formula: N_{Ed} = 178 ≤ 12.5 • (2•14-7) = 263. An interaction calculation is therefore not necessary and a sheet pile wall can be assumed which is only vertically loaded, with:
- a vertical load N_{Ed} = 1,55 (276 - 286 + 125) = 178 kN/m to be absorbed exclusively by the point.
- a vertical load N_{Ed} = 1,55 (276 + 125) = 622 kN/m to be absorbed by the point and the shaft alongside the paint surface.

On the basis of the cone penetration test S09, the following follows at *NAP* -3.00 metres with a cone resistance of q_c = 16 MPa and a ξ4-value of 1.03 for n = 4 cone penetration tests (please note: if calculating point bearing capacity by D-Sheet Piling, then $p_{r;max;poilt}$ = 16 • 0.65 = 10.4 MPa and ξ = 1/1.03 = 0.97):

$$R_{b;d} = \frac{1}{1,03 \cdot 1,2} \cdot 0,023 \cdot 16.000 \cdot 0,62 = 185 \text{ kN/m}$$

$$R_{s;d} = \frac{1}{1,03 \cdot 1,2} \cdot 2 \cdot 1,46 \cdot 0,006 \cdot [8000 \cdot 3 + 9000 \cdot 3] = 723 \text{ kN/m}$$

155

For the first and second approach, the following applies:

- $N_{Ed} = 178 < R_{b;d} = 185$, u.c. = 0.96 (suffices);

- $N_{Ed} = 622 < R_{b;d} + R_{s;d} = 908$, u.c. = 0.69 (also suffices).

On the basis of the two tests above, the vertical balance is met.

Moreover, it follows that the first approach is normative (as is also applied in D-Sheet Pile).

Table 22 Partial material factors for retaining structures.

Point level [m+NAP]	length [m]	W_{soil} [%]	$F_{A;MAX}$ [kN/m]	$M_{Ed;min}$ [kNm/m]	$M_{Ed;max}$ [kNm/m]	$U_{x;max}$ [mm]
-7.0	18	47.6	-298.73	-584.8	325.0	79.8
-6.0	17	53.2	-301.85	-596.5	289.4	-82.8
-5.0	16	60.1	-309.64	-626.1	210.3	-89.2
-4.0	15	67.5	-321.01	-670.5	104	-96.7
-3.0	14	76.0	-330.64	-709.0	19.4	-100.4
-2.0	13	92.0	-332.92	-718.3	8.4	-101.6
-1.0	12	> 100	instable sheet pile wall			

Step 5c: Test pile driving depth functional requirements

No functional requirements as acquitards apply to the determination of the required pile driving depth.

Step 6a up to 6f: Dimensioning calculations

By means of computer programme D-Sheet Pile, the various calculation steps in the dimensioning calculation have been performed, for which a sensitivity analysis has been conducted for the water levels and the values of the subgrade reaction modulus. As a free water level occurs on the low side, calculation steps 6.1 and 6.2 have not been calculated. The results are mentioned in table 23 for both the construction phase with RC1 and the end phase for the sheet pile wall under strain of bending (group C) as well as the end phase for the vertical balance (group B). For the latter, only the maximum anchor force is relevant. In the construction phase, the anchor has been post-tensioned at 100 kN/m.

Table 23 Results dimensioning calculations

Construction phase	Calculation step	M_{Ed} [kNm/m]	$V_{E;max}$ [kN/m]	$F_{A;max}$ [kN/m]	$N_{E;max}$ [kN/m]	W_{soll} [%]	$u_{x;max}$ [mm]
1 (construction phase)	a: 6.1	-	-	-	-	-	-
	b: 6.2	-	-	-	-	-	-
	c: 6.3	280	142	100	64	45	-
	d: 6.4	275	135	100	64	44	-
	e: 6.5	149	71	100	64	27	13
	f: 6.5 x 1.5	179	85	120	77	-	-
2 (final situation group C)	a: 6.1	-	-	-	-	-	-
	b: 6.2	-	-	-	-	-	-
	c: 6.3	710	235	331	212	76	-
	d: 6.4	687	231	325	209	82	-
	e: 6.5	321	130	183	118	43	35
	f: 6.5 x 1.5	385	155	219	141	-	-
3 (final situation group B)	a: 6.1	-	-	-	-	-	-
	b: 6.2	-	-	-	-	-	-
	c: 6.3	-	-	192	124	-	-
	d: 6.4	-	-	192	124	-	-
	e: 6.5	-	-	-	-	-	-
	f: 6.5 x 1.5	-	-	-	-	-	-

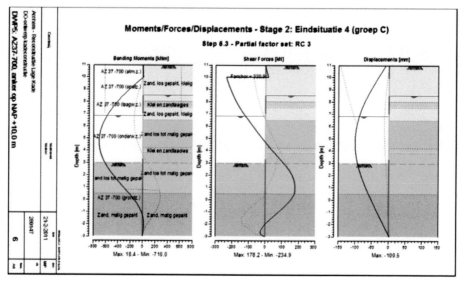

Figure 5-13 Graphic calculation results sheet pile wall calculation phase 2.

Step 7a: Calculate maximum moment of occurrence

The following follows from figure and table 23:
- above *NAP* +7.0 m: maximum moment M_{Ed} = 570 kNm/m (at *NAP* +7.0 m);
- above *NAP* +7.0 m: maximum moment M_{Ed} = 710 kNm/m (at *NAP* +4.7 m).

Step 7b: Calculate moment capacity

The moment capacity has been calculated on the basis of the normative corrosion reductions and the correction factor for the strength of the sheet pile wall material resulting from the higher reliability class.

For the sheet pile wall above *NAP* + 7.0 metres, the following applies:
- bending stiffness: EI_{cor} = 102,204 kNm² /m;
- resistance moment: $W_{el;cor}$ = 1,951 cm³ /m;
- cross-section class: (b/t_f) / ε = 450 / (17-8.6) / $\sqrt{(235/355)}$ = 65.9
  so cross-section class 3;
- correction partial material factor steel: $\gamma_{M;st}$ = 1.0 • 1.066 = 1.066;
- calculation may only be elastic, so $M_{c,Rd}$ = 1,951 • 355 • 10^{-3} / 1.066 = 650 kNm/m;
- test moment: u.c. = 570 / 644 = 0.88 < 1.0 so it suffices.

For the sheet pile wall under *NAP* + 7.0 metres, the following applies:
- bending stiffness: EI_{corr} = 140,647 kNm² /m;
- resistance moment: $W_{el;cor}$ = 2,684 cm³ /m;
- cross-section class: (b/t_f) / ε = 450 / (17-5,0) / $\sqrt{(235/355)}$ = 46.1
  so cross-section class 3;

158

- correction partial material factor steel: $\gamma_{M;st} = 1.0 \cdot 1.066 = 1.066$;
- calculation may only be elastic, so $M_{c,Rd} = 2,684 \cdot 355 \cdot 10^{-3} / 1.066 = 894$ kNm/m;
- test moment: u.c. = 710 / 887 = 0.79 < 1.0 so it suffices.

Step 7c: Determine possible reduction factor as a result of large differences in water pressure

The maximum difference in water level amounts to 1.5 metres and lies far below the difference of 5 metres, so reduction does not apply.

Step 8a: Determine the situations with maximal transverse forces and normal forces

The following follows from figure 5-13 and table 23:
- under *NAP* +7.0 m: maximum moment $M_{Ed} = 710$ kNm/m;
- accompanying maximum transverse force: $V_{Ed} = 235$ kN/m;
- accompanying maximum axial force: $N_{Ed} = 212$ kN/m.

Step 8b: Test strength sheet pile wall by means of material-bound Eurocodes

On the basis of Eurocode 3 [20], the strength of the sheet pile wall was tested for the above normative combination of loads. To test buckling stability, the following applies in the thinnest cross-section with cross-section class 3 on the basis of a buckling length equal to the length of the sheet pile wall between anchor level and point sheet pile wall:

$$N_{cr} = \frac{102204 \cdot 1 \cdot \pi^2}{13} = 77.593 \text{ kN/m}$$

$$\frac{N_{Ed}}{N_{cr}} = 0{,}0027 \leq 0{,}04$$

An extra buckling stability test is therefore unnecessary. For the axial load test (including the correction factor $\gamma_{M;st} = 1.066$), the following applies:

$$N_{pl;Rd} = \frac{0{,}023 \cdot 355000}{1{,}066} = 7.660 \text{ kN/m}$$

$$\frac{N_{Ed}}{N_{pl,Rd}} = 0{,}027 \leq 0{,}10$$

An additional test for the share of normal tension on the total bend tension of the sheet pile wall is therefore unnecessary. A separate test for the transverse tension from the transverse force is also unnecessary. For the testing of the strength of the steel, a moment capacity test is sufficient, which has already been conducted in step 7.

Step 9a: Calculate the anchor force with the 5% upper limit

This calculation step is not relevant, as the stiffness of the anchor is known.

Step 9b: Determine the normative anchor force

The anchors will be placed every 2 double piles, due to which the centre-to-centre distance between two anchors will amount to 2.8 metres. For the dimensioning of the anchors, the following two calculated forces are normative and relevant (including the correction factor $\gamma_{M;st} = 1.066$):
- maximal anchor force in the ULS: $P_{max} = 2.8 \cdot 331 \cdot 1.066 = 985$ kN;
- maximal anchor force in the SLS: $P_{max} = 2.8 \cdot 183 = 512$ kN.

Step 9c: Determine the calculation value of the normative anchor force

As the anchorage is permanent (and the sheet pile wall falls under RC3 and this is also true for RC2 and RC3 for temporary anchors), an assessment of the anchor failure for anchors and walings is obligatory. For both situations, the following calculation values (including the correction factor 1.066 for the reliability index increase) apply:

ultimate limit state:
- $P_{rod;d} = 988 \cdot 1.25 = 1236$ kN;
- $P_{grout;d} = 988 \cdot 1.1 \cdot 1.066 = 1087$ kN.

calamity in case of failure:
- $P_{rod;d} = 1.5 \cdot 512 \cdot 1.0 = 768$ kN;
- $P_{grout;d} = 1.5 \cdot 512 \cdot 1.0 = 768$ kN.

The ultimate limit state is therefore more normative than the calamity situation in case of anchor failure. Moreover, for the testing of the strength material factors equal to 1.0 may be used for calculations, which further increases the difference.

Step 9d: Test strength and bearing capacity anchorage structure

An anchor calculation has been conducted on the basis of cone penetration test S09. For screw injection anchors, the following is assumed:
- limiting q_c on 20 MPa;
- shaft friction factor $\alpha_t = 0.015$;
- calculation diameter equal to $1.5 \cdot D_{drill\ head}$ in conformity with CUR 166 [8.16];
- on all anchor tests, so factor $\xi = 1.0$ and $\gamma_a = 1.25$.

For the anchor rod, the following is applied:
- TITAN anchor type 73/35;
- steel quality: MW450;
- flow force: $R_y = 1355/1.0 = 1355$ kN (normative);

160

- breach force: R_{br} = 1980/1.4 = 1414 kN;
- strength test rod: u.c. = 1236 / 1355 = 0.91 ($\leq$ 1.0 so rod suffices).

On the basis of the test of the tension bearing capacity, the following applies for the pull-out resistance:
- drill head diameter: Ø200 mm;
- calculation diameter: $D_{grout;d}$ = 1.5 • 200 = 300 mm;
- length grout body: 9.0 metres (from *NAP* +1.0 to -4.8 m);
- average cone resistance on the basis of cone penetration test S9: q_c = 12.1 MPa;

$$R_{a;d} = \frac{\pi \cdot 0,03}{1,25 \cdot 1,0} \cdot 0,015 \cdot \sum_{i=1}^{n} (q_{c;i;gem} \cdot 9,0) = 1228 \text{ kN/m}$$

- pull-out resistance test: u.c. = 1087 / 1228 = 0.89 ($\leq$ 1.0 so grout body suffices).

Additional test deep-seated grout body

The top of the grout body lies at *NAP* +1.0 m. From this, the following applies:
- test cover on the grout body: h = 11.0 – 1.0 = 10.0 metres (suffices, because h > 5.0 m);
- test embedding sand: top of consolidated sand layer starts at *NAP* +2.0 metres so at least 1.0 metres under top of (consolidated) sand layer.

Additional testing situation with corrosion

As the anchorage is permanent, it is possibly subjected to corrosion. The anchors are fitted with double corrosion allowance, which means that corrosion is no longer applicable.

Additional test settling soil on the anchor rods

This test is not discussed in this example.

Step 10: Deformation test

The maximum calculated deformation amounts to 35 mm. For permanent sheet pile walls, Dutch department of Infrastructure uses 1/200 of the retaining height, this being 40 mm. This means that it suffices.

Step 11a: Test Kranz stability

Figure 5-14 presents the testing of the Kranz stability. From this an anchor force capacity follows according to Kranz : $R_{Kr;k} = 760$ kN/m. From the test, the following applies (including the correction factor $\gamma_{M;st} = 1.066$):

$$P_{max} \leq \frac{760}{1.5 \cdot 1.066} = 476 \text{ kN/m}$$

As $P_{max} = 330$ kN/m, the above requirement is met.

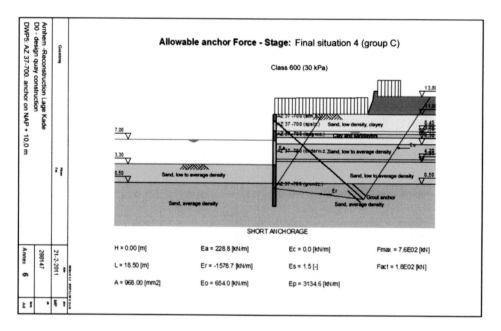

Figure 5-14 Test Kranz stability.

Step 11b: Test overall stability

Figure 5-15 presents the test of the overall stability. The safety factor is calculated on the basis of the calculation value in conformity with table 31 with adjustment of the correction factors $\gamma_{M;corr}$. With the stability factor of 1.31 > 1.0, the stability requirement is met.

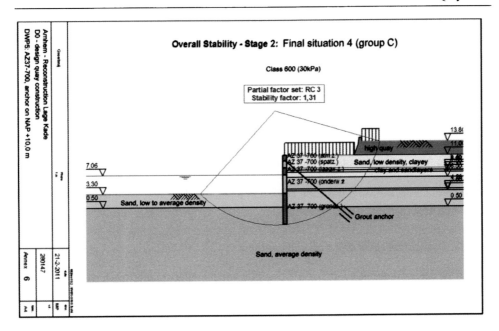

Figure 5-15 Test overall stability phase 2.

Step 11c: Test ground breach

The ground breach is tested in step 5, where the percentage of mobilised soil resistance is calculated. This suffices in the ULS on the basis of a point level at *NAP* -3.0 m. The test of the SLS is not normative and is also not relevant for testing a failure mechanism in terms of breach.

Step 11d: Test failure due to flow

Risk of flow does not apply to this calculation example and has not been tested in further detail.

Step 11e: Test vibration and/or sound effects on the surroundings

The feasibility test of the sheet pile wall is conducted on the basis of NVAF graph HF15 (high frequency vibration, cone resistance 15 MPa), as this has been included in appendix B of CUR 166. The applied sheet pile wall has been drawn in the graph in figure 5-16, that is generated in D-Sheet Pile.

On the basis of the resistance moment of 3,700 cm^3 /m and the sheet pile length of 14 metres, the sheet pile wall is located just inside the damage-free installation area, which is limited by a pile driving resistance (for high frequency vibration) of approximately $F_{eff;soil}$ = 2000 kN. This also follows from the prediction of the vibrating block.

163

Step 11f: Test feasibility sheet pile wall

The testing and dimensioning of the sheet pile wall with regard to the effect of vibration on the surroundings have been conducted separately. In these tests, the effects of vibration on the surroundings with regard to boundary values for damage to adjacent buildings have been considered, in conformity with SBR-A. This concerns:
- the effect on the high quay behind the low quay;
- the effect on buildings behind the quay;
- the effect on the foundation of the bridge across the quay.

Based on this information it has been chosen to place the sheet pile wall under the bridge in a vibration-free manner.

Step 12: Verification sheet pile wall calculations

During the design process, a number of principles and boundary conditions have been altered, due to which a verification calculation has become necessary. The effect of the additional load resulting from settling soil on the anchor rods has been looked at. In the verification calculation, an extra tension force has been applied to the sheet pile wall as extra post-tension force, due to which both the sheet pile wall and the waling are under more strain. The increase of the anchor force in the ultimate limit state has been calculated as a post-tension force of approximately 50 kN/m (leading to $331 + 138/2.8 = 380$ kN/m).

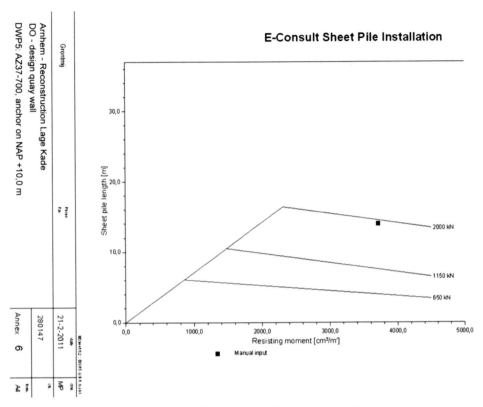

DWP5: AZ37-700, anchor on NAP +10,0 m

DO - design quay wall

Arnhem - Reconstruction Lage Kade

Grontmij

Annex 6

280147

21-2-2011

Figure 5-16 Test feasibility sheet pile wall on the basis of NVAF graph HF15.

For the testing of the strength and deformation of the sheet pile wall (group C), the increase in anchor force leads to a decrease of the bending moment from 710 to 693 kN/m. The deformation remains unaltered.

For the testing of the vertical bearing capacity, a calculation step has been calculated with post-tension force (group B) of $192.4 + 138/2.8 = 241.7$ kN/m. The following applies from the calculation:

- anchor force: $F_{A;max}$ = 241.7 kN/m;
- vertical load active soil pressure: $F_{v;active}$ = 312 kN/m;
- vertical load passive soil pressure: $F_{v;passive}$ = -305 kN/m;
- vertical anchor force: $F_{v;A;max}$ = 241.7 • sin 40° = 155 kN/m;
- total vertical load on sheet pile wall point: N_{Ed} = 1.55 • (312 - 305 + 155) = 251 kN/m.

On the basis of the fixation length, the following applies in conformity with CUR 166:

N_{Ed} = 251 ≤ 12.5 • (2•14-7) = 263, so an interaction calculation is not necessary here and an exclusively vertically loaded sheet pile wall can be assumed. For the first (normative) approach from step 5, a point bearing capacity of 185 kN follows, so from the unity check, it follows that: u.c. = 251/185 = 1.36, and with that the vertical balance would not suffice in the ULS situation.

Also when the results from step 5 are assumed, multiplied with the SLS value of the additional vertical anchor force of $(138/2.8) \cdot \sin 40° = 31$ kN/m, with a load of $31 + 178 = 209$ kN/m, the vertical balance would still be insufficient.

Only if the total load due to settlement is approached as SLS situation (that is, with a load factor of $\gamma_{G;dst;corr} = 1.0$ instead of 1.55), a vertical load of 162 kN/m follows with which the vertical balance is sufficient.

The following solutions can also be chosen for this problem:
- considering the plasticity behaviour of the anchor rod;
- recalculating the reliability index for the sheet pile wall as a result of including additional failure mechanisms of post-settlement on anchor rod;
- sharper recalculation of vertical balance sheet pile wall in the PLAXIS (EEM) programme;
- measures: injection on the tip or 0.5 metre longer sheet pile wall ('cut' method).

Applying the above solutions will lead to the conclusion that, for the sheet pile wall, the vertical balance could be sufficient after all. Elaboration of this has been excluded in this example.

6 Construction

6.1 In general

The appropriate type of contract between client and contractor is important for an optimal risk distribution in the construction phase of a project. This chapter describes which aspects should be taken into account for these contracts, while at the same time focusing on the available tools.

The construction aspects should be included - as much as possible - in the design of the structures. This chapter also provides a summary of specific focus points in construction that should always be taken into account in the design of urban quay walls.

Specific construction aspects and risks are discussed in great detail. In urban areas, environmental factors play a crucial role. These factors will be discussed as well.

6.2 Types of contracts

When talking about a construction process, a client always has a specific result in mind. This result can be described in terms of quality, cost and time. Certain aspects may be more important for the client than others, depending on the particular circumstances.

The client must choose a specific organisation of the construction process, taking the desired result and the focus on particular aspects (quality, costs and time) into account. In doing so, the client must carefully consider the choice of construction organisation.

The construction organisation says something about the distribution, nature and extent of the tasks of the different parties involved in the construction process. The execution of tasks can be arranged in different ways. It is important that the client chooses the tasks on the basis of the objectives he wishes to achieve. Furthermore, the choice of a particular construction organisation must be in line with the decisions made in the context of the requirement specifications.

Requirement specifications and construction organisations are closely linked. Requirement specifications differ from each other in terms of the degree of freedom with regard to solutions they provide to the market. This means that every requirement specification should serve as a mirror of the nature and extent of the influence on the construction process. This to the degree to which the client regards these as suitable for himself and the contacted market parties, given his choice of a particular construction organisation method. Influence on the construction process and freedom (with regard to solutions) are therefore interchangeable concepts.

In fact, the contract type may be added to the aforementioned choices After all, the influence of the construction process combined with the associated freedom offered to the market should have a legislative interpretation in the contractual division of responsibilities and risks.

A client should not only carefully consider the type of contract - especially the nature and content of the contractual conditions - in terms of the desired consistency with the construction organisation and the requirement specification, he should also do this from the perspective of decisions he must make with regard to a number of important aspects in the tendering process.

Aside from this, a well thought-out type of contract in itself is an important part of the process organisation that the client chooses in view of the desired result.

First, because the choice of a particular contractual division of responsibilities and risks is highly dependent upon the number of tenders. If more responsibilities and risks are placed on the market than is apparent from the perspective of the scope of solutions offered to the market, a situation may occur in which the client ultimately pays for value which is, in reality, not created. Second, a well though-out type of contract is important because, with the choice of a particular contractual division of responsibilities and risks, the client determines his future legal obligations. If the client places less responsibilities and risks on the market than expected given the choice of a particular construction organisation and requirement specification, this may lead to a situation in which he is ultimately held responsible for the risks that should be placed on the other party. However, this does not mean that providing a large degree of freedom of solutions to the market will translate in a broad contractual division of responsibilities and risks. In the end, it depends on the goal the client wishes to achieve with the construction process and what his priorities are in this context.

Aside from the aforementioned considerations, a client is well advised to determine the type of contract in a legal context as well. The following aspects should be taken into account:
- Internal policies. In making his decisions - especially in terms of the legal administrative conditions he wishes to include - the client may be restricted by internal policies that prohibit the (unchanged) implementation of certain contractual conditions.
- Tendering legislation limits. Decisions regarding the contract type may be subject to objections from a public tendering law perspective. More specifically, this concerns the application of unbalanced legal-administrative conditions to the contract.

Table 24 provides an overview of the distribution of risk for a given contract. In traditional contracts, the risk responsibilities are mainly born by the client. In integrated contracts, the risks transfer to the contractor.

The Committee on Public Tendering has created the interactive Decision Support System (DSS), which is available via www.leidraadaanbesteden.nl. By asking different questions about a specific project, this tool helps clients make balanced decisions for tenders and contracts. It provides a well-substantiated and reliable decision-making process with pre-determined steps that guarantee legal certainty.

Table 24 Responsibilities for various types of tenders and methods of collaboration [3].

Construction phases	Traditional collaboration			(multi-annual) maintenance concept	Integrated collaboration	
	Director	UAV/RAW	Construction crew	Framework contract	Design & construct	Turnkey
Initiative	Responsibility of client					
Research						
Definition						
ToR						
Preliminary design						
Final design						
Construction design						
Work preparation						
Construction				Responsibility of contractor		
Maintenance						
Framework						
Procurement	Procurement procedure according to current procurement regulation / guidelines					
Construction	UAV	UAV	RVOI/UAV	UAVgc	UAVgc	UAVgc

6.3   Alignment of design and construction

Optimal alignment between design and construction is possible with integrated contracts in which the contractor is responsible for both the design and construction. After all, the design can be fully attuned to the construction options and preferences of the contractor. Through an adequate alignment of design and construction, money and time can be saved. Time can also be saved because tasks in the design and construction phase can be performed in parallel.

However, integrated contracts are not always possible or desirable, see section 6.2. In such cases, the client's designer must - on the basis of past experience and expertise - come up with the best possible implementations for the design.

Table 25 includes points of attention CUR commission C186 "Urban quay walls" that the designer must always take into account in the design of urban quay walls.

Table 25 Points of attention when aligning design and construct.

Point of attention	Potential bottlenecks during construction	Measures during design
Environment objects (buildings, cables and pipes, roads, trees, etc.)	Damage caused by vibration	• Elements need to be applicable in a vibration-free/low fashion
	Damage caused by conflicts between newly designed elements and existing environment objects.	• Timely inventory of existing environmental objects (measurements, KLIC reports, archive research, etc.) and adapt design to this. • Prescribe preemptive intake
	Damage caused by manoeuvring equipment	• Parts near damage-sensitive environment objects need to be applicable with minimal manoeuvring of equipment.
	Damage caused by drainage	• Designed in such a way that no/limited drainage is required (masonry, concrete work and welding as much as possible above waterline) • Controlled drainage, preferably in a cofferdam.
Nearby residents	Construction nuisance (including noise, vibration, etc.)	• Components need to be applicable in a vibration-free/low fashion. • Minimise number of pile driven activities. • Maximise number of prefabricated components (minimising construction time on site).
	Safety of residents during construction	• Designed in such a way that prevents /reduces risky pile driving manoeuvres

Traffic	Obstruction/nuisance to motorway traffic	• Design in such a way that enables construction from the water side. • Maximise number of prefabricated components (minimise construction time on site).
	Obstruction/nuisance to waterway traffic	• Designed in such a way that enables construction from the land side (as much as possible). • Maximise number of prefabricated components (minimise construction time on site).
Transport route	Limit dimensions (height, width, length, draught) and axle load of transport vehicles	• Take the maximum dimensions and weight of the transportable components into account in the design phase.
Available construction space	Limited movement space for equipment	• Components need to be applicable by equipment that can easily manoeuvre within the available space.
	Limited allowable construction load	• Components need to be installed by equipment that meet the maximum construction load (possibly using pressure relieving measures).
	Limited space for auxiliary constructs	• Taking into account available auxiliary constructs for the limited space.
Water levels	No/limited drainage possible due to risk of setting.	• Design in such a way that no/limited drainage is required during construction (masonry, concrete work and welding as much as possible above waterline) • Controlled drainage, preferably in a cofferdam.
	Construct elements under/near water level	• Masonry, concrete and welding work (wherever possible) above water level
Delivery time	Materials not deliverable on time	• Take into account delivery times of materials.

6.4 Construction aspects

Construction work on quay walls in urban areas requires additional attention because of the complexity of environmental factors. There is little room for construction activities in densely built environments with often busy roads. Table 26 lists the points of attention for a specific construction activity that should always be taken into account.

Table 26 Points of attention in construction.

No.	Component	Activities	Specific points of attention during construct
1	Groundwork	Excavation	• Cables and pipes • Flora & Fauna • Archaeological objects • NBE (Non Blown Explosives) • Stability surrounding objects • Soil quality
		Temporary soil storage	• Space and permit for depot • Soil transportation through urban areas (by road or water)
		Replenishment	• Soil degree of compaction • Soil quality • Damage caused by vibration degree of compaction
2	Dredging	Excavation waterway near quay	• Flora & Fauna • Maximum allowable dredging depth (due to quay stability) • Soil quality
		Disposal of sludge/soil	• Soil transportation through urban areas (by road or water) • Processor of sludge/soil

3	Demolition	Removing masonry	• Flora & Fauna • Debris in water
		Removal anchorage	• Quay stability
		Removal concrete structure (gravity wall/ capping beam)	• Noise • Debris in water • Vibrations
		Removal sheet pile wall	• Effect of vibration on environment objects (settlement) • Vibration nuisance (persons, sensitive devices) • NBE (Non Blown Explosives) • Manoeuvre space for equipment
		Removal column foundation	• Effect of vibration on environment objects • Vibration nuisance (persons, sensitive devices) • NBE (Non Blown Explosives) • Manoeuvre space for equipment • Risk of pile fracture • Settlement as a result of spaces in soil

4	Pile driving work	Installing of column foundation	• Effect of vibration on environment objects • Vibration nuisance (persons, sensitive devices) • NBE (Non Blown Explosives) • Cables and pipelines • Manoeuvre space for equipment • Obstacles in the subsoil
		Installing sheet pile wall	• Effect of vibration on environment objects • Vibration nuisance (persons, sensitive equipment) • NBE (Non Blown Explosives) • Cables and pipes • Manoeuvre space for equipment • Obstacles in the subsoil
5	Anchorage	Installing grout anchors	• Archaeological objects • NBE (Non Blown Explosives) • Cables and pipelines • Existing (underground) environment objects • Manoeuvre space for equipment • Property boundaries throughout the anchorage area
6	Concrete work	Installing gravity wall/ capping beam	• Water levels • Outside temperature • Auxiliary structures (see 10) • Concrete supply (see 9)
		Installing gravity wall/ prefab capping beam	• Transport prefab elements (see 9) • Handling prefab elements on location
7	Steel work	Applying anchor plates, reinforced concrete cap and/or quay equipment (bollards, staircases, etc.)	• Water levels • Conservation (restriction under water)
8	Masonry work	Bricklaying prewall	• Water levels • Outside temperature

9	Transportation	Supply and disposal of equipment and materials	• By water or land • Limitations dimensions and weights • Planning permits • Traffic measures/diversions
		Transshipment at construction site	• Capacity hoist facilitation versus available space • Space for storage
10	Auxiliary structures	Cofferdam	• Position stamping • Drainage • Stability surrounding objects • See 4 Installing sheet pile wall
		Concrete shuttering	• Water levels
		Trench shuttering	• Stability surrounding objects

6.5 Construction risks

Construction activities on quay walls in urban areas come with certain risks. These risks are listed in table 27, including possible control measures. Several of these measures will be explained in more detail in later sections.

Table 27 Construction risks

1	Damage to environment objects (buildings, cables and pipelines, roads, trees, etc.)	Vibration caused by foundation work (pile driving and vibrations)	• See section 6.6.3 Preventive precautions
		Maneuvering equipment	• See section 6.6.3 Preventive precautions • Use of protective structures (spreader plates, dragline mats, etc.)
2	Unacceptable nuisance for surrounding residents	Noise construction activities exceeds maximum allowable value	• See section 6.6.3 Preventive precautions
		Local residents insufficiently informed about construction activities	• See section 6.6.4 Communication with Third Parties

3	Unacceptable nuisance road traffic / fairway	Insufficient communication with fairway / road operators and authorities	• See section 6.6.4 Communication with Third Parties
4	Mishaps / accidents in public areas	Unsafe situations in public areas due to construction activities	• See section 6.6.3 Precautions public spaces
5	Damage to the Environment, Flora and Fauna	Insufficient attention paid to Environment, Flora and Fauna during construction activities	• See section 6.6.3 Precautions for Damage to Environment, Flora and Fauna
6	Instability of quay wall during construction phase	Exceeding the maximum allowable construction load behind the quay wall	• Inventory of upper load on equipment behind quay wall per construction phase • Determine maximum allowable upper load behind quay wall per construction phase • If occurring upper load is larger than allowable limit, adjust construction methods (construction from water side, other application of equipment, adjustment of construction planning, etc.)
		Waterbed in front of quay wall dredged too deeply	• Determine maximum allowable dredge depth (including dredging tolerances)
		Incorrect construction planning	• Continue using construction planning during construction phase as used in design phase • Align construction planning between design and construction

7	Construction components cannot be installed	Obstacles in the subsoil	• See section 6.6.3 Measures for Cables and pipelines • Archival research into former and existing structures in the subsoil • Research on location (interviews, trial trenches, surveying, etc.)
		Impossible to transfer components to the construction site	• Identify constraints along transportation routes . • Take these constraints into account in the design phase (alignment design with execution, see section 6.3) • Consultation and arrangements with road operators and authorities, see section 6.6.4 Communication with Third Parties
		Impossible to store components at construction site	• Identify possibilities for storage at construction site (required hoisting capacity in relation to available space) • Take into account storage possibilities in the design phase (align design with construction) • Align hoisting arrangements with available space
		No equipment available to install components	• Inventory of available equipment (noise output, available space, capacity, etc.). • Take design into account regarding available equipment (align design with construction)
		Insufficient space for equipment at construction site to install components	• Inventory of available space for equipment • Take usage of materials into account in design phase (alignment design with construction)
		Construction site suspect to NBE	• Historical research • Exploratory research • Further research • Disposal of detected NBE

177

6.6   Interface and environment

6.6.1   In general

An urban area is a dynamic environment in which individuals and organisations have their own needs, desires and interests. Stakeholders become more demanding and wish to exert influence on organisations, decisions and construction projects. If stakeholder interests are not sufficiently respected or taken into account in projects, chances are these stakeholders will turn against the project, causing unnecessary delays.

The identification of various interests and potential implications for stakeholders helps to determine the correct communication methods and nuisance-reducing measures.

6.6.2   Nuisance caused by method of construction

As a result of the construction work, there is a risk of possible nuisance in surrounding areas. Examples include persons, road vehicles, houses, industrial buildings, infrastructure works, public areas and flora and fauna that are exposed to noise, vibrations, damage caused by materials or potential unsafe situations. To reduce this nuisance, the construction activities that may cause nuisance must be determined and managed at the earliest possible stage. The following sections list potential measures that can prevent environmental nuisance during the construction activities.

6.6.3   Preventive precautions

During the construction activities, there is a risk of damage to flora, fauna or the environment in general, to adjacent buildings, roads, bridges, quay walls, banks or public facilities (hereinafter referred to as objects). Damage can be caused by:
- work activities that involve vibrations or where cables and pipes are affected;
- collision with ships and work boats;
- hoisting and other movements on or near objects;
- construction methods.

Furthermore, the work shall be executed in urban areas where the 'construction site' is freely accessible to passers-by, tourists or even playing children. This presents additional safety risks. In addition, in urban areas it is likely that the immediate environment views the construction activities as a nuisance. The following sections will elaborate on how these risks can be managed through appropriate measures.

Precautions public spaces The construction activities mostly take place in public spaces where bystanders have direct access to the construction site. The bystanders are unaware of the dangers and risks associated with the work itself. To adequately protect these people, the following measures may be taken:

- Prepare demolition and construction safety plan for demolition and construction activities;
- prevent erosion or washing out of the bank;
- execute work activities - with the exception of light earthwork - from the water side;
- physically mark the construction site with guide beacons, traffic barriers, shields and/or screens (in consultation with municipalities and road authorities);
- close off location by means of physical barriers;
- apply alarm circuits (trespassing, entering construction site is life threatening);
- shield and seal machines, tools and other equipment when leaving the construction site;
- moor vessels in appropriate locations so that they cannot be accessed from the bank;
- escape routes, fire hydrants and other public facilities should be available and accessible;
- shut down electrical installations after working hours so that unauthorised persons cannot access them, with the exception of required lighting;
- storage of materials and unloading equipment on the water by means of pontoons or on a special site in the vicinity of the construction site, to prevent abuse and vandalism;
- storing of construction components and hoisting of materials according to applicable guidelines (see below);
- place pedestrian containers so they can pass safely when hoisting long components;
- 24/7 monitoring of construction sites;
- good communication with the neighbouring area (residents, road authorities, etc.).

Specifically with regard to hoisting and other manoeuvres:

- Hoisting of materials according to NVAF guidelines Foundation Works along motorways and infrastructural projects with increased safety risk for public environment.
- If necessary, perform all work from the water side, supply and removal of waste via waterways and no storage of materials on the bank. Nearby loading/unloading quay walls may be used for storage, loading and unloading of materials.
- Use of tested and certified equipment. The foundation machines have a TCVT Certificate of Approval. Stability calculations are available for the pontoons. CUR committee C186 'urban quay walls'.
- Run test for foundation equipment before starting work.
- Use qualified personnel hoisting work - VVL certified hitcher (safe moving and attaching of loads), TCVT certified machinist (monitoring certification vertical transport), diploma foundation worker and Head sail license for boaters.
- Use of vessels and pontoons with spud poles during hoisting work for no discomfort because of waves.
- Do not perform hoisting operations in adverse weather conditions (from wind force 6 Beaufort).

- Use of approved and suitable hoisting equipment (piling clamp, safety hooks, etc.).

Measures for cables and pipes

An important aspect in urban areas are the cables and pipes that run through the area of construction. These must be identified and marked before commencement of work activities. A KLIC report should be made in the design phase, so this aspect can be taken into account as early as possible.

To be certain about the exact position or (unknown) location of cables and pipes, these must be identified before commencement of the work activities using a lance and possibly divers assistance on the water side or by digging trenches on the land side. The exact position of the cables and pipes must be marked on the construction site as highlighted on the design drawings.

Prior to construction and with appropriate frequency, the contractor, authorities of the cables and pipes and, if necessary, the client, must plan and come to an agreement on the following aspects:
- conflicts, risks and supervision;
- the proposed construction methodology;
- project agreements in terms of construction methodology, management and detailed solutions such as underpasses or connections.

To prevent or minimise damage to cables and pipes, the following measures may be taken:
- performing a KLIC report before start of the work and creation of a Cables and Pipes dosser;
- organisation of coordination meetings and reaching project agreements with authorities;
- moving cables and pipes;
- designing quay walls with safety margin for cables and pipes;
- identifying cables and pipes, marking by means of crane ship, lance and possible divers assistance or digging trenches on the land side;
- observance of authorities' regulations, WION and CROW guideline Careful excavation process [2] to prevent damage, inform all involved employees;
- use of surveyor when installing components in quay wall within critical safety margins of cables and pipes;
- extra instructions for personnel and continuous monitoring of construction in case of critical safety margins.

Precautions for vibration and deformation damage

Because of the various work activities, local residents, businesses and objects may be exposed to vibrations. This may cause nuisance and possible damage to buildings and businesses in the vicinity of the construction site.

The likelihood of nuisance or (deformation) damage (subsidence or settlement) to objects or buildings depends highly on the type of work. For example, the installation of sheet pile walls, the excavation of soil and the delivery of heavy equipment (construction vehicles) comes with increased risks. The following measures can be taken to reduce vibration nuisance:

- limiting the amount of vibration-related work activities that are performed simultaneously and in the same area of influence;
- adjustment of construction methods, such as use of heavy vibrators/rams (e.g. vibration frequency, pressure, power);
- use of a high-frequency vibrator with a 'variable time', which helps reduce noise pollution in the vicinity of the work;
- positional engineering recordings, vibration and deformation measurements, see below.

By performing vibration measurements, positional recordings and deformation measurements, deformation damage can be detected in time and preventive/corrective measures can be taken. In doing so, the following method may be applied:

- Determine vibration prognosis and the construction site's area of influence taking sounding, the design and equipment into account to decrease the risk of deformation or vibration damage. Per object, the chance of deformation and vibration damage (in accordance with SBR guideline A [34]), the vibration nuisance for persons (in accordance with SBR guideline B[35]) and the vibration nuisance for sensitive equipment (in accordance with SBR guideline C [36]) is determined.
- On the basis of the area of influence and environment analysis, determine which objects (properties, roads, listed buildings) require positional recording and vibration and deformation measurements. If an area of influence covers parts of a block of houses, it is advisable to assume a high risk for the full block. This to prevent discussions about inclusion or exclusion of individual houses in a housing block.
- Creation of a monitoring plan in accordance with SBR guidelines or CUR 223 [9] prior to the construction activities for the measurement and monitoring of construction pits. The aim is to prevent vibration and deformation damage. During monitoring any damage can be timely identified and measures can be taken to prevent further damage. Within this monitoring plan, three types of monitoring can be distinguished: (1) architectural engineering recordings (0-situation), (2) vibration measurements and (3) deformation and height measurements. The monitoring plan should include maximum deformation and vibration values (limit values) per category (1, 2 or 3) in accordance with SBR guidelines or CUR 223. As long as the limit values are not exceeded, there is virtually no chance of vibration damage (incl. cosmetic damage) as a result of the construction activities (chance of damage according to SBR guideline < 1%);
- Determine whether project-related insurances cover any damage. If not, take out supplementary insurance to cover unforeseen damage. However, first examine guidelines for insurances.

Precautions noise nuisance

Neighbouring or other noise-sensitive objects may be affected by noise during the construction and demolition activities. Construction and demolition noise are temporary occurrences that are highly dependent upon local circumstances. To a certain degree, the neighbouring area should expect a temporary inconvenience. However, if this inconvenience goes beyond what is socially acceptable, the municipality may impose further regulations to limit the noise pollution. To determine and limit the degree of noise pollution, the following measures can be taken:

- use of double work sets to reduce duration of noise pollution without an additional noise intensity. The intensity of noise is relatively higher when using two sets instead of one, but the construction time is halved;
- expected noise to environment, measure in accordance with instructions and calculate industrial noise and on the basis of the source power equipment and duration of construction activities, calculate and assess circular construction noise;
- if noise on facades exceeds the applicable local limit, take noise abatement measures, such as equipping the waling with plastic foam, installing silencers on equipment and/ or using mobile sound barriers;
- discuss expected noise pollution and noise-reducing measures with all relevant stakeholders (especially locals and offices);
- only perform work activities during the day (Monday to Friday from 7am to 7pm);
- use a high frequency vibrator with a 'variable time', which limits the vibration and noise pollution in the vicinity of the construction site.

Precautions for damage to environment, flora and fauna

In the Netherlands, the protection of about 500 different species of plants and animals is regulated by the Flora and Fauna Act. The Flora and Fauna Act aims to protect and preserve wild plants and animals. The following information is copied from the Code of Conduct in the Flora and Fauna Act [15].

This protection is included in the prohibitions contained in Articles 8 to 13. The law recognises the possibility for an exemption to these prohibitions under certain conditions. Under the Decree exempting protected animal and plant species (Exemption Decree) exemption from the prohibitions (Article 8 - 12) is provided for certain activities as long as activities are performed within the Code of Conduct approved by the Minister of Agriculture. To this end, various public bodies, including local municipalities, have set up their own code of conduct (see: www.hetlnvloket.nl). This leads to a reduction of administrative tasks and procedure times, because one does not need to apply for an exemption under the Flora and Fauna Act as often. The Code of Conduct also provides clarity about the protective measures to be taken.

A Code of Conduct can only be approved as long as it guarantees that activities are performed with care. The approved Code of Conduct subsequently forms a condition to obtain an

Exemption Decree. The concept of 'performing with care' is an important element of this Code of Conduct.

Performing with care implies that the activity or operation poses no material impact on protected species and that, prior to execution of the activities, all reasonable measures are taken to avoid damage to the protected species.

Subsequent (mitigating) measures can be taken to ensure non/minimal disturbance, injury to or death of the protected flora and fauna.
- identify existing flora and fauna;
- identify possible mitigating measures on the basis of and work according to the applicable Code of Conduct so that no permits and exemptions are required;
- creation of work plan flora and fauna by ecologist in which the activities are aligned with the existing flora and fauna and the mitigating measures to be taken, such as planning/phasing of activities outside of sensitive periods, temporarily remove (flora) or catch and set free in different location (fauna);
- if mitigating measures are not possible, apply for exemption Flora and Fauna Act and work according to exemption guidelines;
- before commencement work, have ecologist release a work trajectory;
- execution of work outside of breeding period birds or prevent birds from making nests (mow in time);
- have trees that obstruct the work activities cut by a specialised gardening company and only cut them in extreme cases (and, if necessary, request logging permit);
- use of tree sheathing for protection of trees in the vicinity of the construction site;
- never use hazardous substances that may pollute the (aquatic) environment;
- have an ecologist continuously monitor the construction activities.

Even with direct measures in place to protect flora and fauna, contamination of harbour water or the soil may still occur as a result of the work activities. This contamination may have a direct impact on the flora, fauna and environment near the construction location. Table 28 provides an overview of risks and indirect measures to protect the flora, fauna and environment.

6.6.4 Communication with Third Parties

Nuisance is often a subjective perception and can be minimised through proper and timely communication. To avoid stagnation of the project, there should be a good relationship between the client, contractor, licensing authorities, local residents, businesses and other stakeholders in the immediate vicinity of the construction site. Correct and timely communication is crucial in achieving this. This requires adequate collaboration between the client and contractor.

It is recommended that the client and contractor jointly create an information and communication plan that includes information on the methods of consultation, frequency and parties involved. It should also state what information is communicated and when this is communicated between the environment, client and contractor.

The nuisance to shipping traffic, road traffic, businesses and local residents can be minimised in advance through efficient public communication about the nature, duration and possible disruptions of the construction activities.

Below is a list of recommendations for adequate communication with the public:
- Client and contractor shall jointly create an information and communication plan that includes information on the methods of consultation, frequency and parties involved. It should also state what information is communicated and when this is communicated between the environment, client and contractor.
- Client and contractor regularly consult about the division of roles regarding communication with local residents and businesses. In these talks, the method and content of the communication is finalised and specialised contact points are appointed.
- Timely and before start of the work activities, inform residents and businesses by letter about the nature, duration and times during which the work is executed in the vicinity of their homes/offices, what type of nuisance they can expect and what measures need to be taken to prevent nuisance, damage and risks.
- Organise information evenings, for which all stakeholders are invited to be informed about the construction times and the expected nuisance caused by the work.
- Use the website of the client and contractor to inform interested parties about the nature, duration and nuisance caused by the work.
- Use Social Media (Twitter, Facebook…) and allow interested parties to actively ask questions about the project.
- Establish an information centre in the immediate vicinity of the construction site.
- Establish a procedure for the purpose of dealing with complaints, comments and/or questions from local residents.

Traffic safety on the urban roads requires extra attention. To guarantee this, the road authorities should - prior to the construction activities - be informed about:
- planning and duration of the construction activities, where and when;
- the expected nuisance per nuisance category and its duration;
- required workspace and keeping the waterway sufficiently accessible;
- traffic measures and restrictions (signage, accessibility facilities, boat speed limitation, use VHF radio, and so on);
- execution methodology near flood defences;
- issue notice to skippers (BAS notification);
- creation traffic plan and waterway management plan;
- applying for the necessary permits and exemptions (road regulation, BPR);
- determination type and location of construction and traffic signage;
- placement of advertisements;
- preparing emergency scenario plan.

Table 28 Risks and measures for environmental protection.

Risk or danger	Cause	Measures to be taken
Pollution of harbour water or emission of substances	Water polluted by dust particles/pieces Proliferation of light products Marine litter Odour nuisance Transport movements on dry surfaces	• Assess activities on the basis of "decree discharge outside facilities" and, if necessary, apply for exemption (demolition, installation of anchors, welding, grinding above water level) • Draft a work plan including measures to minimise pollution • When working on soil, always comply with regulations in BRL7000 (avoid smearing and spreading of soil) • Perform construction under favourable weather conditions (no wind/rain/storm/hail) to minimise spreading • Prevent removed/installed components from falling at all times • Apply catching utilities (nets/sheets/foils) • Use of suction equipment or suction bag in the event of dust • Minimise dust particles/sawdust by manually or by using fire equipment to dismantle (bolt) connections • Storage of light products in mesh cages, install ballasts or secure with straps • Immediately clean waste and residues and cover to prevent spread • Immediate removal of material in the water
Oil on surface of water	Damage caused by burst of hydraulic conduits, junctions or pumps	• Sufficiently shield components • If possible, install hose rupture security on equipment • Check leak tightness before starting construction activities • Annual control and inspection of equipment • Have absorbents and oil booms available and, if necessary, use them • In case of leakage, halt all activities, stop leakage and remove contamination, in consultation with client

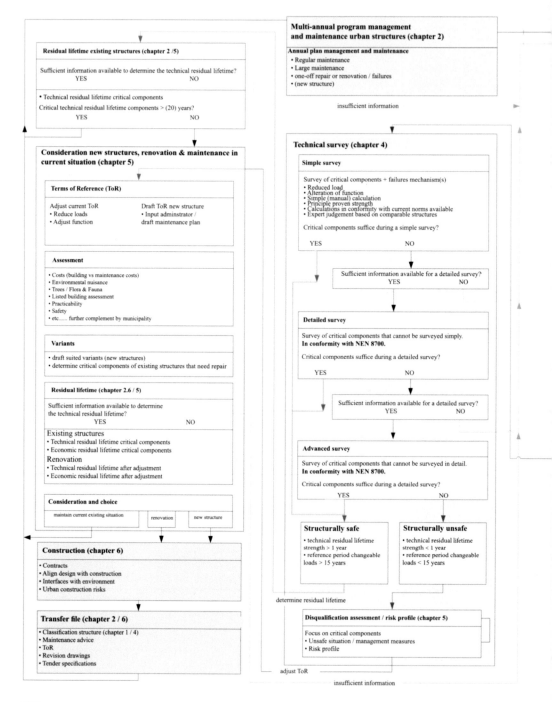

Residual lifetime existing structures (chapter 2 /5)

Sufficient information available to determine the technical residual lifetime?
YES   NO

• Technical residual lifetime critical components
Critical technical residual lifetime components > (20) years?
YES   NO

Consideration new structures, renovation & maintenance in current situation (chapter 5)

Terms of Reference (ToR)

Adjust current ToR   Draft ToR new structure
• Reduce loads   • Input adminstrator /
• Adjust function   draft maintenance plan

Assessment

• Costs (building vs maintenance costs)
• Environmental nuisance
• Trees / Flora & Fauna
• Listed building assessment
• Practicability
• Safety
• etc...... further complement by municipality

Variants

• draft suited variants (new structures)
• determine critical components of existing structures that need repair

Residual lifetime (chapter 2.6 / 5)

Sufficient information available to determine
the technical residual lifetime?
YES   NO

Existing structures
• Technical residual lifetime critical components
• Economic residual lifetime critical components
Renovation
• Technical residual lifetime after adjustment
• Economic residual lifetime after adjustment

Consideration and choice

maintain current existing situation | renovation | new structure

Construction (chapter 6)

• Contracts
• Align design with construction
• Interfaces with environment
• Urban construction risks

Transfer file (chapter 2 / 6)

• Classification structure (chapter 1 / 4)
• Maintenance advice
• ToR
• Revision drawings
• Tender specifications

Multi-annual program management and maintenance urban structures (chapter 2)

Annual plan management and maintenance
• Regular maintenance
• Large maintenance
• one-off repair or renovation / failures
• (new structure)

insufficient information

Technical survey (chapter 4)

Simple survey

Survey of critical components + failures mechanism(s)
• Reduced load
• Alteration of function
• Simple (manual) calculation
• Principle proven strength
• Calculations in conformity with current norms available
• Expert judgement based on comparable structures

Critical components suffice during a simple survey?

YES   NO

Sufficient information available for a detailed survey?
YES   NO

Detailed survey

Survey of critical components that cannot be surveyed simply.
In conformity with NEN 8700.

Critical components suffice during a detailed survey?

YES   NO

Sufficient information available for a detailed survey?
YES   NO

Advanced survey

Survey of critical components that cannot be surveyed in detail.
In conformity with NEN 8700.

Critical components suffice during a detailed survey?

YES   NO

Structurally safe

• technical residual lifetime
strength > 1 year
• reference period changeable
loads > 15 years

Structurally unsafe

• technical residual lifetime
strength < 1 year
• reference period changeable
loads < 15 years

determine residual lifetime

Disqualification assessment / risk profile (chapter 5)

Focus on critical components
• Unsafe situation / management measures
• Risk profile

adjust ToR

insufficient information

Appendix 1 Flowchart survey method

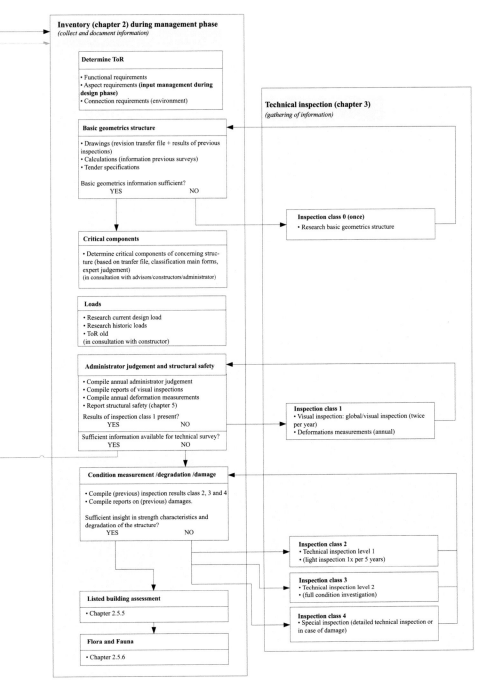

Figuur B1-1.

Appendix 2
Valuation chart listed building value

A. Cultural and historical values
1. Importance of the object/property as a particular expression of (a) cultural, social, economic, administrative, mental and/or other social development(s).
2. Importance of the object/property as a particular expression of (a) technical, production-technical, functional and/or typological development(s).
3. Importance of the object/property because of innovative value or groundbreaking character.
4. Particular importance of the object/property for the history of architecture and/or engineering.
5. Particular importance of the object/property for the work of a master builder, architect, engineer, manufacturer or company.
6. Importance of the object/property because of educational or museum values.

B. Beauty
1. Importance of the object/property because of the high aesthetic and architectural qualities of the design.
2. Importance of the object/property because of the unique use of materials and/or special ornamentation.
3. Importance of the object/property because of the unique relationship between exterior and interior (parts) or between components of the object/property.
4. Importance of the object/property because of the high architectural qualities of its constituent parts.

C. Ensemble value
1. Special significance of the object/property as an (essential) part of a larger cultural-historical, architecture-historical, town-planning or landscape importance.
2. Special significance of the property because of the allotment/furnishing/facilities.

D. Authenticity and recognisability
1. Importance of the object/property because of the architectural authenticity/recognisability of exes/interior.
2. Importance of the object/property because of the structure's authenticity.

E. Rarity and representativeness

1. Importance of the object/property because of architecture-historical, structural, typological and functional rarity, possibly linked to a particular age.
2. Importance of the object/property because it is representative for its time, place or region, the design, layout and construction.

Appendix 3
Partial factors - new structures

Note 1: The classification in the following tables is according to EN 1990, EN 1997 and BS 8700: consequence classes CC1, CC2 and CC3 correspond to respective reliability classes RC1, RC2 and RC3.

Note 2: The values for the partial material factors from the tables below are applicable if a high value of the material parameter is unfavourable. If a low value of the parameter is favourable, a value of 1.0 must be taken for the partial factor.

Table 29 Partial factors retaining wall type 1: retaining wall on spread foundations (new structure).

Application and source	Parameter	Symbol	Value CC1 (β=3.3)	Value CC2 (β=3.8)	Value CC3 (β=4.3)
Structural load Eurocode 7, table A.3 column A1 (group B)	Permanent load, unfavourable[1]	γ_G	1.2	1.35	1.5
	Permanent load, unfavourable[2]	$\gamma_G \times \xi$	1.1	1.2	1.3
	Permanent load, favourable	$\gamma_{G;stb}$	0.9	0.9	0.9
	Variable load, unfavourable	$\gamma_{Q;dst}$	1.35	1.5	1.65
	Variable load, favourable	$\gamma_{Q;dst}$	0	0	0
Geotechnical load Eurocode 7, table A.3 column A2 "Other" (group C)	Permanent load, unfavourable	γ_G	1.0	1.0	1.0
	Permanent load, favourable	$\gamma_{G;stb}$	1.0	1.0	1.0
	Variable load, unfavourable	$\gamma_{Q;dst}$	1.2	1.3	1.4
	Variable load, favourable	$\gamma_{Q;dst}$	0	0	0

Application and source	Parameter	Symbol	Value CC1 (β=3.3)	Value CC2 (β=3.8)	Value CC3 (β=4.3)
Material factors bearing capacity Eurocode 7, table A.4a column "Retaining wall"	Angle of internal friction	$\gamma_{\varphi'}$ [3]	1.2	1.2	1.2
	Effective cohesion	$\gamma_{c'}$	1.5	1.5	1.5
	Undrained shear strength	γ_{cu}	1.5	1.5	1.5
	Compressive prism strength	γ_{qu}	1.5	1.5	1.5
	Density	γ_{γ}	1.1	1.1	1.1
Material factors general stability Eurocode 7, table A.4a column "General stability"	Angle of internal friction	$\gamma_{\varphi'}$ [3]	1.2	1.25	1.3
	Effective cohesion	$\gamma_{c'}$	1.3	1.45	1.6
	Undrained shear strength	γ_{cu}	1.5	1.75	2.0
	Compressive prism strength	γ_{qu}	1.5	1.75	2.0
	Density	γ_{γ}	1.0	1.0	1.0
Material factors deformations Eurocode 7, table A.4c column M2	Compression indices, unfavourable	$\gamma_{cc}, \gamma_{c\alpha}, \gamma_{csw}$ [4]	0.8	0.8	0.8
	Compression indices, favourable	$\gamma_{cc}, \gamma_{c\alpha}, \gamma_{csw}$ [4]	1.0	1.0	1.0
	Compression constants, unfavourable	γ_{cp}, γ_{cs} [5]	1.0	1.0	1.0
	Compression constants, favourable	γ_{cp}, γ_{cs} [5]	1.3	1.3	1.3

1) Only applicable on small variable loads

2) Only when share of variable load is large (in conformity with NEN-EN 1990/NB, apply $\xi = 0,89$)

3) Factor concerns tan φ'.

4) Factors concern $C_c/(1+e_0)$ and C_α and $C_{sw}/(1+e_0)$, and a-, b-, c-isotachs and modified compression indices λ^*, k^* and μ^*.

5) Factors also concern C_p' and C_s'.

Tabel 30 Partial factors retaining wall type 2 and 3: retaining wall on pile foundation (new structure).

Application	Parameter	Symbol	Value CC1 (β=3.3)	Value CC2 (β=3.8)	Value CC3 (β=4.3)
Structural and geotechnical loads Eurocode 7, table A.3 column A1 (group B)	Permanent load, unfavourable[1]	γ_G	1.2	1.35	1.5
	Permanent load, unfavourable[2]	$\gamma_G \times \xi$	1.1	1.2	1.1
	Permanent load, favourable	$\gamma_{G;stb}$	0.9	0.9	0.9
	Variable load, unfavourable	$\gamma_{Q;dst}$	1.35	1.5	1.65
	Variable load, favourable	$\gamma_{Q;dst}$	0	0	0
Material factors pile bearing capacity Eurocode 7, tables A.4a column "Foundations" A.6/A.7/A.8 column R3-c	Cone resistance by probing	γ_b	1.2	1.2	1.2
	Shaft resistance by probing (pressure)	γ_s	1.2	1.2	1.2
	Total burst pressure by probing	γ_t	1.2	1.2	1.2
	Shaft resistance by probing (tensile)	$\gamma_{s;t}$	1.35[4]	1.35[4]	1.35[4]
	Angle of internal friction	γ_φ[3]	1.15	1.15	1.15
	Effective cohesion	$\gamma_c{'}$	1.6	1.6	1.6
	Density	γ_γ	1.1	1.1	1.1
Material factors general stability Eurocode 7 table A.4a column "General stability"	Angle of internal friction	$\gamma_\varphi{'}$[3]	1.2	1.25	1.3
	Effective cohesion	$\gamma_c{'}$	1.3	1.45	1.6
	Undrained shear strength	γ_{cu}	1.1	1.75	2.0
	Compressive prism strength	γ_{qu}	1.5	1.75	2.0
	Density	γ_γ	1.0	1.1	1.0
Material factors deformations Eurocode 7 table A.4c column M2	Load/settlement behaviour, unfavourable	$\gamma_{settlement\ s}$[5]	1.3	1.3	1.3
	Load/settlement behaviour, favourable	$\gamma_{settlement\ s}$[5]	1.0	1.0	1.0
	Soil compression factor, E-moduli, unfavourable	γ_{kh}, γ_E	1.0	1.0	1.0
	Soil compression factor, E-moduli, favourable	γ_{kh}, γ_E	1.3	1.3	1.3

1) Only applicable on small variable loads
2) Only when share of variable load is large (in conformity with NEN-EN 1990/NB, apply ξ = 0,89)
3) Factor concerns tan φ'
4) For anchor piles specific factors are applicable
5) Multiplier on calculated settlement from the serviceability limit state for the calculation of settlement in the limit state.

Table 31 Partial factors retaining wall type 4: sheet pile wall structures (new structure).

Application	Parameter	Symbol	Value RC1 (β=3.3)	Value RC2 (β=3.8)	Value RC3 (β=4.3)
Constructive and geotechnical loads (Eurocode 7, table A.3 column A2 "Sheet pile wall" (group C)	Permanent load, unfavourable[1]	γ_G	1.0	1.0	1.0
	Permanent load, unfavourable[2]	$\gamma_{G;stb}$	1.0	1.0	1.0
	Variable load, unfavourable	$\gamma_{Q;dst}$	1.0	1.1	1.25
	Variable load, favourable	$\gamma_{Q;dst}$	0	0	0
Loads vertical bearing sheet pile wall (being foundation) Eurocode 7, table A.3 column A1 (group B)	Permanent load, unfavourable[1]	γ_G	1.2	1.35	1.5
	Permanent load, unfavourable[2]	$\gamma_G \times \xi$	1.1	1.2	1.1
	Permanent load, favourable	$\gamma_{G;stb}$	0.9	0.9	0.9
	Variable load, unfavourable	$\gamma_{Q;dst}$	1.35	1.5	1.65
	Variable load, favourable	$\gamma_{Q;dst}$	0	0	0
Loads on anchorage and related constructions	Load on anchor tie rod	$\gamma_{f;a}$	1.25	1.25	1.25
	Load on grout body, anchor wall	$\gamma_{f;a}$	1.10	1.10	1.10
Material factors earth pressures on sheet pile wall Eurocode 7, table A.4b column "Sheet pile wall"	Angle of internal friction	$\gamma_\varphi{}'^{3)}$	1.15	1.175	1.2
	Effective cohesion	$\gamma_c{}'$	1.15	1.25	1.1
	Undrained shear strength	γ_{cu}	1.5	1.6	1.65
	Compressive prism strength	γ_{qu}	1.5	1.6	1.65
	Density	γ_γ	1.0	1.0	1.0
	Soil compression factor, E-moduli, unfavourable	γ_{kh}, γ_E	1.0	1.0	1.0
	Soil compression factor, E-moduli, favourable	γ_{kh}, γ_E	1.3	1.3	1.3
Material factors vertical bearing capacity Eurocode 7, tables A4.a column "Foundations" A.6/A.7/A.8 column R3-c	Cone resistance by probing	γ_b	1.2	1.2	1.2
	Shaft resistance by probing (pressure)	γ_s	1.2	1.2	1.2
	Angle of internal friction	$\gamma_\varphi{}'^{3)}$	1.15	1.15	1.15
	Effective cohesion	$\gamma_c{}'$	1.6	1.6	1.6
	Density	γ_γ	1.1	1.1	1.1

Material factors general stability Eurocode 7, table A.4a column "General stability"	Angle of internal friction	γ_φ'[3]	1.2	1.25	1.3
	Effective cohesion	γ_c'	1.3	1.45	1.6
	Undrained shear strength	γ_{cu}	1.5	1.75	2.0
	Compression prism strength	γ_{qu}	1.5	1.75	2.0
	Density	γ_γ	1.0	1.0	1.0
Material factors anchor resistance Eurocode 7 table A.12	Permanent anchor, no control test	$\gamma_{a;p}$	1.25	1.25	1.25
	Permanent anchor, control test	$\gamma_{a;p}$	1.35	1.35	1.35
Material factors sheet pile wall materials CUR 166	Steel sheet pile wall	γ_M	1.0	1.0	1.0
	Concrete sheet pile wall	γ_M	1.1	1.1	1.1
	Wooden sheet pile wall	γ_M	1.2	1.2	1.2
Material factors deformations Eurocode 7, table A.4c column M2	Load/settlement behaviour, unfavourable	$\gamma_{settlement\ s}$[4]	1.3	1.3	1.3
	Load/settlement behaviour, favourable	$\gamma_{settlement\ s}$[4]	1.0	1.0	1.0
	Soil compression factor, E-moduli, unfavourable	γ_{kh}, γ_E	1.0	1.0	1.0
	Soil compression factor, E-moduli, favourable	γ_{kh}, γ_E	1.3	1.3	1.3

1) Only applicable on small variable loads

2) Only when share of variable load is large (in conformity with NEN-EN 1990/NB, apply ξ = 0,89)

3) Factor concerns tan φ'.

4) Multiplier on calculated settlement from the serviceability limit state for the calculation of settlement in the limit state.

197

Appendix 4
Partial factors - repair/renovation

Note 1: The classification in the following tables is according to EN 1990, EN 1997 and BS 8700: consequence classes CC1, CC2 and CC3.

Note 2: The values for the partial material factors from the tables below are applicable if a high value of the material parameter is unfavourable. If a low value of the parameter is favourable, a value of 1.0 must be taken for the partial factor.

Note 3: When partial material factors for new structures are the same, in accordance with the Eurocode, Annex B, these will remain unchanged for both renovation and disqualification (existing structure) [1].

Note 4: The tables list the origin (substantiation) of the different partial factors.

Table 32 Partial factors retaining wall type 1: retaining wall on spread foundation (renovation).

Application and source	Parameter	Symbol	Value CC1 $(\beta=2.8)^{6)}$	Value CC2 $(\beta=3.3)$	Value CC3 $(\beta=3.8)$
Structural loads, article Vrouwenvelder et. al., table 4[1]	Permanent load, unfavourable[1]	γ_G	1.15	1.3	1.4
	Permanent load, unfavourable[2]	$\gamma_G \times \xi$	1.05	1.15	1.25
	Permanent load, favourable	$\gamma_{G;stb}$	0.9	0.9	0.9
	Variable load, unfavourable	$\gamma_{Q;dst}$	1.1	1.3	1.5
	Variable load, favourable	$\gamma_{Q;dst}$	0	0	0
Geotechnical loads Eurocode 7, table A.3 column A2 "Other" (group C) and CUR 166, table 3.7	Permanent load, unfavourable	γ_G	1.0	1.0	1.0
	Permanent load, favourable	$\gamma_{G;stb}$	1.0	1.0	1.0
	Variable load, unfavourable	$\gamma_{Q;dst}$	(1.0)	1.2	1.3
	Variable load, favourable	$\gamma_{Q;dst}$	0	0	0

Application and source	Parameter	Symbol	Value CC1 (β=2.8)[6]	Value CC2 (β=3.3)	Value CC3 (β=3.8)
Material factors bearing capacity Eurocode 7, table A.4a column "Retaining wall"	Angle of internal friction	$\gamma_{\varphi'}$ [3]	1.2	1.2	1.2
	Effective cohesion	$\gamma_{c'}$	1.5	1.5	1.5
	Undrained shear strength	γ_{cu}	1.5	1.5	1.5
	Compressive prism strength	γ_{qu}	1.5	1.5	1.5
	Density	γ_{γ}	1.1	1.1	1.1
Material factors general stability Eurocode 7, table A.4a column "General stability" and CUR 162, table 3.6	Angle of internal friction	$\gamma_{\varphi'}$ [3]	(1.1)	1.2	1.25
	Effective cohesion	$\gamma_{c'}$	(1.1)	1.3	1.45
	Undrained shear strength	γ_{cu}	(1.2)	1.5	1.75
	Compressive prism strength	γ_{qu}	(1.2)	1.5	1.75
	Density	γ_{γ}	(1.0)	1.0	1.0
Material factors deformations Eurocode 7, table A.4c column M2	Compression indices, unfavourable	γ_{cc}, $\gamma_{c\alpha}$, γ_{csw} [4]	0.8	0.8	0.8
	Compression indices, favourable	γ_{cc}, $\gamma_{c\alpha}$, γ_{csw} [4]	1.0	1.0	1.0
	Compression constants, unfavourable	γ_{cp}, γ_{cs} [5]	1.0	1.0	1.0
	Compression constants, favourable	γ_{cp}, γ_{cs} [5]	1.3	1.3	1.3

1) Only applicable on small variable loads

2) Only when share of variable load is large (in conformity with NEN-EN 1990/NB, apply $\xi = 0{,}89$)

3) Factor concerns tan φ'.

4) Factors concern $C_c/(1+e_0)$ and C_α and $C_{sw}/(1+e_0)$, and a-, b-, c-isotachs and modified compression indices λ^*, k^* and μ^*.

5) Factors also concern C_p' and C_s'.

6) Values in brackets are in accordance with $\beta = 2{,}5 \sim 2{,}6$ in conformity with CUR-class I.

Table 33 Partial factors retaining wall type 2 and 3: retaining wall on pile foundation (renovation).

Application	Parameter	Symbol	Value CC1 (β=2.8)[6]	Value CC2 (β=3.3)	Value CC3 (β=3.8)
Structural loads, article Vrouwenvelder et. al., table 4[0]	Permanent load, unfavourable[1]	γ_G	1.15	1.3	1.4
	Permanent load, unfavourable[2]	$\gamma_G \times \xi$	1.05	1.15	1.25
	Permanent load, favourable	$\gamma_{G;stb}$	0.9	0.9	0.9
	Variable load, unfavourable	$\gamma_{Q;dst}$	1.1	1.3	1.5
	Variable load, favourable	$\gamma_{Q;dst}$	0	0	0
Material factors pile bearing capacity Eurocode 7, tables A.4a column "Foundations" A.6/ A.7/A.8 column R3-c	Cone resistance by probing	γ_b	1.2	1.2	1.2
	Shaft resistance by probing (pressure)	γ_s	1.2	1.2	1.2
	Total burst pressure by probing	γ_t	1.2	1.2	1.2
	Shaft resistance by probing (tensile)	$\gamma_{s;t}$	1.35[4]	1.35[4]	1.35[4]
	Angle of internal friction	γ_φ'[3]	1.15	1.15	1.15
	Effective cohesion	γ_c'	1.6	1.6	1.6
	Density	γ_γ	1.1	1.1	1.1
Material factors general stability Eurocode 7, table A.4a column "General stability" and CUR 162, table 3.6	Angle of internal friction	γ_φ'[3]	(1.1)	1.2	1.25
	Effective cohesion	γ_c'	(1.1)	1.3	1.45
	Undrained shear strength	γ_{cu}	(1.2)	1.5	1.75
	Compressive prism strength	γ_{qu}	(1.2)	1.5	1.75
	Density	γ_γ	(1.0)	1.0	1.0

Application	Parameter		Value CC1 $(\beta=2.8)$ [6]	Value CC2 $(\beta=3.3)$	Value CC3 $(\beta=3.8)$
Material factors deformations Eurocode 7, table A.4c column M2	Load/settlement behaviour, unfavourable	$\gamma_{settlement\,s}$ [5]	1.3	1.3	1.3
	Load/settlement behaviour, favourable	$\gamma_{settlement\,s}$ [5]	1.0	1.0	1.0
	Soil compression factor, E-moduli, unfavourable	γ_{kh}, γ_E	1.0	1.0	1.0
	Soil compression factor, E-moduli, favourable	γ_{kh}, γ_E	1.3	1.3	1.3

1) Only applicable on small variable loads

2) Only when share of variable load is large (in conformity with NEN-EN 1990/NB, apply $\xi = 0,89$)

3) Factor concerns tan φ'

4) For anchor piles specific factors are applicable

5) Multiplier on calculated settlement from the serviceability limit state for the calculation of settlement in the limit state.

6) Values in brackets are in accordance with $\beta = 2,5 \sim 2,6$ in conformity with CUR-class I

Table 34 Partial factors retaining wall type 4: sheet pile wall structures (renovation).

Application	Parameter	Symbol	Value CC1 (β=2.8)[5]	Value CC2 (β=3.3)	Value CC3 (β=3.8)
Structural and geotechnical loads (Eurocode 7, table A.3 column A2 "Sheet pile wall" and CUR 166 table 3.7	Permanent load, unfavourable[1]	γ_G	(1.0)	1.0	1.0
	Permanent load, unfavourable[2]	$\gamma_{G;stb}$	(1.0)	1.0	1.0
	Variable load, unfavourable	$\gamma_{Q;dst}$	(1.0)	1.0	1.1
	Variable load, favourable	$\gamma_{Q;dst}$	0	0	0
Structural loads article Vrouwenvelder et.al., table 4 (0)	Permanent load, unfavourable[1]	γ_G	1.15	1.3	1.4
	Permanent load, unfavourable[2]	$\gamma_G \times \xi$	1.05	1.15	1.25
	Permanent load, favourable	$\gamma_{G;stb}$	0.9	0.9	0.9
	Variable load, unfavourable	$\gamma_{Q'dst}$	1.1	1.3	1.5
	Variable load, favourable	$\gamma_{Q;dst}$	0	0	0
Loads on anchorage and related constructions	Load on anchor tie rod	$\gamma_{f;a}$	1.25	1.25	1.25
	Load on grout body, anchor wall	$\gamma_{f;a}$	1.10	1.10	1.10
Material factors earth pressures on sheet pile wall Eurocode 7, table A.4b column "Sheet pile wall", CUR 166, table 3.7 and CUR 162, table 3.6	Angle of internal friction	$\gamma_\varphi'^{3}$	(1.05)	1.15	1.175
	Effective cohesion	γ_c'	(1.05)	1.15	1.25
	Undrained shear strength	γ_{cu}	(1.2)	1.5	1.6
	Compressive prism strength	γ_{qu}	(1.2)	1.5	1.6
	Density	γ_γ	(1.0)	1.0	1.0
	Soil compression factor, E-moduli, unfavourable	γ_{kh}, γ_E	(1.0)	1.0	1.0
	Soil compression factor, E-moduli, favourable	γ_{kh}, γ_E	(1.3)	1.3	1.3

203

Material factors vertical bearing capacity Eurocode 7, tables A4.a column "Foundations" A.6/A.7/A.8 column R3-c	Cone resistance by probing	γ_b	1.2	1.2	1.2
	Shaft resistance by probing (pressure)	γ_s	1.2	1.2	1.2
	Angle of internal friction	γ_φ [3]	1.15	1.15	1.15
	Effective cohesion	γ_c'	1.6	1.6	1.6
	Density	γ_γ	1.1	1.1	1.1
Material factors general stability Eurocode 7, table A.4a column "General stability" and CUR 162, table 3.6	Angle of internal friction	γ_φ [3]	(1.1)	1.2	1.15
	Effective cohesion	γ_c'	(1.1)	1.3	1.45
	Undrained shear strength	γ_{cu}	(1.2)	1.5	1.75
	Compression prism strength	γ_{qu}	(1.2)	1.5	1.75
	Density	γ_γ	(1.0)	1.0	1.0
Material factors anchor resistance Eurocode 7 table A.12	Permanent anchor, no control test	$\gamma_{a;p}$	1.25	1.25	1.25
	Permanent anchor, control test	$\gamma_{a;p}$	1.35	1.35	1.35
Material factors sheet pile wall materials CUR 166	Steel sheet pile wall	γ_M	1.0	1.0	1.0
	Concrete sheet pile wall	γ_M	1.1	1.1	1.1
	Wooden sheet pile wall	γ_M	1.2	1.2	1.2
Material factors deformations Eurocode 7, table A.4c column M2	Load/settlement behaviour, unfavourable	$\gamma_{settlement\ s}$ [4]	1.3	1.3	1.3
	Load/settlement behaviour, favourable	$\gamma_{settlement\ s}$ [4]	1.0	1.0	1.0
	Soil compression factor, E-moduli, unfavourable	γ_{kh}, γ_E	1.0	1.0	1.0
	Soil compression factor, E-moduli, favourable	γ_{kh}, γ_E	1.3	1.3	1.3

1) Only applicable on small variable loads

2) Only when share of variable load is large (in conformity with NEN-EN 1990/NB, apply $\xi = 0{,}89$)

3) Factor concerns $\tan \varphi'$.

4) Multiplier on calculated settlement from the serviceability limit state for the calculation of settlement in the limit state.

5) Values in brackets are in accordance with $\beta = 2{,}5 \sim 2{,}6$ in conformity with CUR-class I

Appendix 5
Partial factors - existing structure

Note 1: The classification in the following tables is according to EN 1990, EN 1997 and BS 8700: consequence classes CC1, CC2 and CC3.

Note 2: The values for the partial material factors from the tables below are applicable if a high value of the material parameter is unfavourable. If a low value of the parameter is favourable, a value of 1.0 must be taken for the partial factor.

Note 3: When partial material factors for new structures are the same, in accordance with the Eurocode, Annex B, these will remain unchanged for both renovation and disqualification (existing structure) [1].

Note 4: The tables list the origin (substantiation) of the different partial factors.

Note 5: Given the fact that CC1 with $\beta = 1.8$ at a reject level corresponds to a safety level in the usability limit state, the material factors are equated to 1.0 (actual strength) [1].

Table 35 Partial factors retaining wall type 1: retaining wall on spread foundation (disqualification).

Application	Parameter	Symbol	Value CC1 ($\beta=1.8$)[6]	Value CC2 ($\beta=2.5$)	Value CC3 ($\beta=3.3$)
Structural loads, article Vrouwenvelder et. al., table 4[0]	Permanent load, unfavourable[1]	γ_G	1.1	1.2	1.3
	Permanent load, unfavourable[2]	$\gamma_G \times \xi$	1.0	1.1	1.2
	Permanent load, favourable	$\gamma_{G;stb}$	0.9	0.9	0.9
	Variable load, unfavourable	$\gamma_{Q;dst}$	1.05	1.15	1.3
	Variable load, favourable	$\gamma_{Q;dst}$	0	0	0
Geotechnical loads Eurocode 7, table A.3 column A2 "Other" (group C) and CUR 166, table 3.7	Permanent load, unfavourable	γ_G	1.0	1.0	1.0
	Permanent load, favourable	$\gamma_{G;stb}$	1.0	1.0	1.0
	Variable load, unfavourable	$\gamma_{Q;dst}$	1.0	(1.0)	1.2
	Variable load, favourable	$\gamma_{Q;dst}$	0	0	0

Material factors	Description	Symbol			
Material factors bearing capacity Eurocode 7, table A.4a column "Retaining wall"	Angle of internal friction	$\gamma_{\varphi'}$ [3]	1.0	1.2	1.2
	Effective cohesion	$\gamma_{c'}$	1.0	1.5	1.5
	Undrained shear strength	γ_{cu}	1.0	1.5	1.5
	Compressive prism strength	γ_{qu}	1.0	1.5	1.5
	Density	γ_{γ}	1.0	1.1	1.1
Material factors general stability Eurocode 7, table A.4a column "General stability" and CUR 162, table 3.6	Angle of internal friction	$\gamma_{\varphi'}$ [3]	1.0	(1.1)	1.2
	Effective cohesion	$\gamma_{c'}$	1.0	(1.1)	1.3
	Undrained shear strength	γ_{cu}	1.0	(1.2)	1.5
	Compressive prism strength	γ_{qu}	1.0	(1.2)	1.5
	Density	γ_{γ}	1.0	(1.0)	1.0
Material factors deformations Eurocode 7, table A.4c column M2	Compression indices, unfavourable	$\gamma_{cc}, \gamma_{c\alpha}, \gamma_{csw}$ [4]	1.0	0.8	0.8
	Compression indices, favourable	$\gamma_{cc}, \gamma_{c\alpha}, \gamma_{csw}$ [4]	1.0	1.0	1.0
	Compression constants, unfavourable	γ_{cp}, γ_{cs} [5]	1.0	1.0	1.0
	Compression constants, favourable	γ_{cp}, γ_{cs} [5]	1.0	1.3	1.3

1) Only applicable on small variable loads

2) Only when share of variable load is large (in conformity with NEN-EN 1990/NB, apply $\xi = 0,89$)

3) Factor concerns $\tan \varphi'$.

4) Factors concern $C_c/(1+e_0)$ and C_α and $C_{sw}/(1+e_0)$, and a-, b-, c-isotachs and modified compression indices λ^*, k^* and μ^*.

5) Factors also concern C_p' and C_s'.

6) Values in brackets are in accordance with $\beta = 2,5 \sim 2,6$ in conformity with CUR-class I.

Table 36 Partial factors retaining wall type 2 and 3: retaining wall on pile foundation (disqualification).

Application	Parameter	Symbol	Value CC1 (β=1.8)	Value CC2 (β=2.5)[6]	Value CC3 (β=3.3)
Structural loads, article Vrouwenvelder et. al., table 4[0]	Permanent load, unfavourable[1]	γ_G	1.1	1.2	1.3
	Permanent load, unfavourable[2]	$\gamma_G \times \xi$	1.0	1.1	1.2
	Permanent load, favourable	$\gamma_{G;stb}$	0.9	0.9	0.9
	Variable load, unfavourable	$\gamma_{Q;dst}$	1.05	1.15	1.3
	Variable load, favourable	$\gamma_{Q;dst}$	0	0	0
Material factors pile bearing capacity Eurocode 7, tables A.4a column "Foundations" A.6/A.7/A.8 column R3-c	Cone resistance by probing	γ_b	1.0	1.2	1.2
	Shaft resistance by probing (pressure)	γ_s	1.0	1.2	1.2
	Total burst pressure by probing	γ_t	1.0	1.2	1.2
	Shaft resistance by probing (tensile)	$\gamma_{s;t}$	1.0	1.35[4]	1.35[4]
	Angle of internal friction	γ_φ [3]	1.0	1.15	1.15
	Effective cohesion	γ_c'	1.0	1.6	1.6
	Density	γ_γ	1.0	1.1	1.1
Material factors general stability Eurocode 7, table A.4a column "General stability" and CUR 162, table 3.6	Angle of internal friction	γ_φ [3]	1.0	(1.1)	1.2
	Effective cohesion	γ_c'	1.0	(1.1)	1.3
	Undrained shear strength	γ_{cu}	1.0	(1.2)	1.5
	Compressive prism strength	γ_{qu}	1.0	(1.2)	1.5
	Density	γ_γ	1.0	(1.0)	1.0

Application	Parameter	Symbol			
Material factors deformations Eurocode 7, table A.4c column M2	Load/settlement behaviour, unfavourable	$\gamma_{settlement\ s}$ [5]	1.0	1.3	1.3
	Load/settlement behaviour, favourable	$\gamma_{settlement\ s}$ [5]	1.0	1.0	1.0
	Soil compression factor, E-moduli, unfavourable	γ_{kh}, γ_E	1.0	1.0	1.0
	Soil compression factor, E-moduli, favourable	γ_{kh}, γ_E	1.0	1.3	1.3

1) Only applicable on small variable loads
2) Only when share of variable load is large (in conformity with NEN-EN 1990/NB, apply $\xi = 0{,}89$)
3) Factor concerns $\tan \varphi'$
4) For anchor piles specific factors are applicable
5) Multiplier on calculated settlement from the serviceability limit state for the calculation of settlement in the limit state.
6) Values in brackets are in accordance with $\beta = 2{,}5 \sim 2{,}6$ in conformity with CUR-class I

Table 37 Partial factors retaining wall type 4: sheet pile wall structures (disqualification).

Application	Parameter	Symbol	Value SLS ($\beta=1.8$)	Value CUR-I ($\beta=2.5$)[5]	Value CC2 ($\beta=3.3$)
Structural and geotechnical loads (Eurocode 7, table A.3 column A2 "Sheet pile wall" (group C)	Permanent load, unfavourable[1]	γ_G	1.0	1.0	1.0
	Permanent load, unfavourable[2]	$\gamma_{G;stb}$	1.0	1.0	1.0
	Variable load, unfavourable	$\gamma_{Q;dst}$	1.0	1.0	1.0
	Variable load, favourable	$\gamma_{Q;dst}$	0	0	0
Structural loads article Vrouwenvelder et.al., table 4(0)	Permanent load, unfavourable[1]	γ_G	1.1	1.2	1.3
	Permanent load unfavourable[2]	$\gamma_G \times \xi$	1.0	1.1	1.2
	Permanent load favourable	$\gamma_{G;stb}$	0.9	0.9	0.9
	Variable load, unfavourable	$\gamma_{Q;dst}$	1.05	1.15	1.3
	Variable load, favourable	$\gamma_{Q;dst}$	0	0	0
Loads on anchorage and related constructions	Load on anchor tie rod	$\gamma_{f;a}$	1.0	1.25	1.25
	Load on grout body, anchor wall	$\gamma_{f;a}$	1.0	1.10	1.10

Material factors earth pressures on sheet pile wall Eurocode 7, table A.4b column "Sheet pile wall", CUR 166, table 3.7 and CUR 162, table 3.6	Angle of internal friction	γ_φ '3)	1.0	(1.05)	1.15
	Effective cohesion	γ_c'	1.0	(1.05)	1.15
	Undrained shear strength	γ_{cu}	1.0	(1.2)	1.5
	Compressive prism strength	γ_{qu}	1.0	(1.2)	1.5
	Density	γ_γ	1.0	(1.0)	1.0
	Soil compression factor, E-moduli, unfavourable	γ_{kh}, γ_E	1.0	(1.0)	1.0
	Soil compression factor, E-moduli, favourable	γ_{kh}, γ_E	1.0	(1.3)	1.3
Material factors vertical bearing capacity Eurocode 7, tables A.4a column "Foundations" A.6/A.7/A.8 column R3-c	Cone resistance by probing	γ_b	1.0	1.2	1.2
	Shaft resistance by probing (pressure)	γ_s	1.0	1.2	1.2
	Angle of internal friction	γ_φ '3)	1.0	1.15	1.15
	Effective cohesion	γ_c'	1.0	1.6	1.6
	Density	γ_γ	1.0	1.1	1.1
Material factors general stability Eurocode 7, table A.4a column "General stability" and CUR 162, table 3.6	Angle of internal friction	γ_φ '3)	1.0	(1.1)	1.2
	Effective cohesion	γ_c'	1.0	(1.1)	1.3
	Undrained shear strength	γ_{cu}	1.0	(1.2)	1.5
	Compression prism strength	γ_{qu}	1.0	(1.2)	1.5
	Density	γ_γ	1.0	(1.0)	1.0
Material factors anchor resistance Eurocode 7 table A.12	Permanent anchor, no control test	$\gamma_{a;p}$	1.0	1.25	1.25
	Permanent anchor, control test	$\gamma_{a;p}$	1.0	1.35	1.35
Material factors sheet pile wall materials CUR 166	Steel sheet pile wall	γ_M	1.0	1.0	1.0
	Concrete sheet pile wall	γ_M	1.0	1.1	1.1
	Wooden sheet pile wall	γ_M	1.0	1.2	1.2

Material factors deformations Eurocode 7, table A.4c column M2	Load/settlement behaviour, unfavourable	$\gamma_{\text{settlement s}}$ [4]	1.0	1.3	1.3
	Load/settlement behaviour, favourable	$\gamma_{\text{settlement s}}$ [4]	1.0	1.0	1.0
	Soil compression factor, E-moduli, unfavourable	γ_{kh}, γ_E	1.0	1.0	1.0
	Soil compression factor, E-moduli, favourable	γ_{kh}, γ_E	1.0	1.3	1.3

1) Only applicable on small variable loads

2) Only when share of variable load is large (in conformity with NEN-EN 1990/NB, apply $\xi = 0,89$)

3) Factor concerns $\tan \varphi'$.

4) Multiplier on calculated settlement from the serviceability limit state (SLS) for the calculation of settlement in the limit state (ULS).

5) Values in brackets are in accordance with $\beta = 2,8$. See also table B3.1 and B3.2.

Appendix 6 Literature

1. 'Beschrijving van handel en nijverheid in Nederland', part 1, p .219, N.V. Boekhandel 's Gravenhage, s.d.
2. CROW, 'Graafschade voorkomen aan kabels en leidingen – Richtlijn zorgvuldig graafproces', Ede, 2008.
3. CROW, 'Uniforme Administratieve Voorwaarden voor geïntegreerde contractvormen', Ede, 2005.
4. CUR-Aanbeveling 72, 'Inspectie en onderzoek van betonconstructies', CUR Gouda, 2011.
5. CUR-Aanbeveling 73, 'Stabiliteit van steenconstructies', CUR Gouda, 2000.
6. CUR-rapport 162, 'Construeren met grond, Grondconstructies op en in weinig draagkrachtige en sterk samendrukbare ondergrond', CUR, Gouda, 2001.
7. CUR-rapport 166, 'Damwandconstructies' Part 1 and 2, CUR, Gouda, 2005.
8. SBRCURnet publication 211E, 'Quay Walls' Second edition, Rotterdam, 2013.
9. CUR-rapport 223, 'Richtlijn meten en monitoren van bouwputten voor kwaliteits-en risicomanagement', CUR, Gouda, 2010.
10. CUR-rapport 236, 'Ankerpalen', CUR, Gouda, 2011.
11. F3O-CUR-SBR, 'Richtlijn Onderzoek en beoordeling van houten paalfunderingen onder gebouwen', CUR-SBR publicatie, 2012.
12. Jorissen, A.J.M., 'Modificatiefactor vochtgehalte en duurbelasting', Technische Hout-documentatie A 4/2 – 010210, Centrum Hout, Almere, 1995
13. Klaassen, R.K.W.M. & B.S. Overeem, 'Factors that influence the speed of bacterial wood degradation', Journal of cultural heritage, 2012.
14. Klaassen, R.K.W.M., 'Aantasting onder water in palen onder gebouwen en waterbouwkundige constructies', 8ste Nationale houten Heipalendag 17 januari, 2013.
15. Ministerie van Infrastructuur en Milieu, Gedragscode Flora- en faunawet, Bestemd voor bestendig beheer en onderhoud en kleinschalige ruimtelijke inrichting of ontwikkeling, 2010.
16. Ministerie van Verkeer en Waterstaat, Voorschrift Toetsen op Veiligheid Primaire Waterkeringen, 2007.
17. NEN-EN1990, Eurocode – Grondslagen van het constructief ontwerp, 2002.
18. NEN-EN 1991-2, Eurocode 1: Belastingen op constructies – Part 2: Verkeersbelasting op bruggen, 2003.
19. NEN-EN 1991-4, Eurocode 1: Belastingen op constructies – Part 1-4: Algemene belastingen – Windbelastingen, 2007.
20. NEN-EN 1993-5, Eurocode 3: Ontwerp en berekening van staalconstructies – Part 5: Palen en damwanden, 2008.
21. NEN-EN 1995-1-1, Eurocode 5: Ontwerp en berekening van houtconstructies, 2005.

22. NEN-EN 338, 'Hout voor constructieve toepassingen' – Sterkteklassen, 2009.
23. NEN-EN-ISO 4628-2, 'Verven en vernissen – Beoordeling van de kwaliteitsafbraak van verflagen – Aanduiding van de intensiteit, hoeveelheid en omvang van algemeen voorkomende gebreken' – Part 2: 'Beoordeling van de mate van blaarvorming', 2003.
24. NEN-EN-ISO 4628-3, 'Verven en vernissen – Beoordeling van de kwaliteitsafname van verflagen – Aanduiding van de hoeveelheid en omvang van gebreken en van de intensiteit van veranderingen' – Part 3: 'Beoordeling van de mate van roestvorming', 2003.
25. NEN-EN-ISO 4628-4, 'Verven en vernissen – Beoordeling van de kwaliteitsafbraak van verflagen – Aanduiding van de intensiteit, hoeveelheid en omvang van algemeen voorkomende gebreken' – Part 4: 'Beoordeling van de mate van barstvorming', 2003.
26. NEN-EN-ISO 4628-5, 'Verven en vernissen – Beoordeling van de kwaliteitsafbraak van verflagen – Aanduiding van de intensiteit, hoeveelheid en omvang van algemeen voorkomende gebreken' – Part 5: 'Aanduiding van de mate van afbladderen', 2003.
27. NEN 2767, 'Conditiemeting' – Part 1: Methodiek, 2011.
28. NEN 3237, 'Cement en kalk – Chemisch onderzoek', 1959.
29. NEN 5491, 'Kwaliteitseisen voor hout (KVH 2010)
 - Heipalen' - Europees naaldhout, 2010.
30. NEN 5968, 'Beton en mortel – Bepaling van de druksterkte van proefstukken', 1988.
31. NEN 8700, 'Beoordeling van de constructieve veiligheid van een bestaand bouwwerk bij verbouw en afkeuren' – Grondslagen, 2011.
32. NEN 9997-1, 'Geotechnisch ontwerp van constructies' – Samenstelling van NEN- EN 1997-1, NEN-EN 1997-1 N/B Nationale bijlage en NEN 9097-1 Aanvullingsnorm bij NEN-EN 1997-1, 2011.
33. Rijkswaterstaat Waterdienst,
 'Technisch Rapport Actuele Sterkte van Dijken', Delft, 2009.
34. SBR richtlijn trillingen, Part A – 'Schade aan gebouwen,
 Meet –en beoordelingsrichtlijn', Rotterdam, 2002.
35. SBR richtlijn trillingen, Part B – 'Hinder voor personen in gebouwen,
 Meet –en beoordelingsrichtlijn', Rotterdam, 2002.
36. SBR richtlijn trillingen, Part C – 'Storing aan apparatuur,
 Meet –en beoordelingsrichtlijn', Rotterdam, 2002.
37. Technische Adviescommissie Waterkeringen, Leidraad Kunstwerken, TAW, 2003.
38. Magazine: PT Civiele Techniek, nr. 2, mei 1989.
39. Van Prooijen, L.A., 'De invoer van Rijns hout per vlot 1650 – 1795,
 In economisch en sociaal-historisch jaarboek', part 53, p.30-79, 1990.
40. Vrouwenvelder, T., Steenbergen, R., Scholten, N., "Veiligheidsfilosofie bestaande bouw, Toepassing en interpretatie NEN 8700"
 Cement magazine issuenumber 4, year 2012.

In general:

• De Gijt, J.G., A History of Quay Walls. Delft, 2010. ISBN: 9789059810327